HELP FROM ABOVE

BOOK TWO

"What Lies Above the Clouds"

DAVID ALAN ARNOLD

Help From Above: Book Two
Copyright © 2018 by David Alan Arnold All rights reserved.
First Edition: October 2018

Cover layout and Interior design: Streetlight Graphics

ISBN: 978-1-7321387-3-5

No part of this book may be reproduced, scanned, or distributed in any printed or electronic form without permission. Please do not participate in or encourage piracy of copyrighted materials in violation of the author's rights. Thank you for respecting the hard work of this author.

CONTENTS

Introduction	1
The Hard Way	5
Quitting Time	7
"We Can't Tell You. It's a Secret…"	9
Weekly Meetings (How I Met My Neighbors)	14
Monday Night Football	22
Signs	26
Garbage	30
Storybook	32
They're All Criminals	40
Gangsters and School Buses	41
Dr. Melville	45
My Prayer	48
"There Are Others…"	50
"Our Hands Are Tied…"	52
The News	54
On My Knees	61
"You Don't Know What You're Doing!"	62
The Story at Storybook Inn	67
No Show	74
"You Should Make a Documentary…"	81
The Pancake Movie	85

Another School Bus Stop	87
Blackmail	98
Lou Rothman	101
Facebook	103
Home Sweet Home	105
"We Have to Stop!"	107
Night Stalkers	109
Last Man Standing	112
Cease and Desist	117
Street Gang	122
Muddy Waters (I'm Not Leaving)	128
Strange Disappearance	131
"They'll Kill You…"	135
Help from New York City	136
Boots on the Ground in Sky Forest	138
Charles Ricciardi	144
Murder, Inc.	147
A Recipe for Disaster	149
Beware of Dog	152
Escape from Storybook Inn	154
The Third World	157
For Some, It Is Already Too Late…	159
Nathan Crawford	161
A Message from Storybook Inn	164
"Don't do it Dave. They'll Kill You…"	174
I'm Doing it Anyway	175
Needle in a Haystack	179
In Harm's Way	183
Toys	186

Scout Mission	191
Mother's Day	194
Robert Larivee	198
Wing and a Prayer	200
Falling	204
"Treat Them Like Family…"	207
Light up the Darkness	209
"We Know Where You Live…"	212
"I Got a Weird Feeling."	219
Janice's Role	220
Memorial Day	222
Airborne	224
Pancake Movie	232
Searching Above the Clouds	235
CSI	242
"I'm Surrounded by Assassins…"	243
Barely Ready	245
Right Place, Wrong Time	249
First Find	251
Crime Scene	254
"Your Son is Loved."	260
The Valley of the Shadow of Death	264
If the Shoe Fits…	270
Angel Flight	272
No Vacancy	279
Help From Above	282
Robert's Plan	287
Snakes	289
Into the Valley of the Shadow of Death	291

The Road to Nowhere	307
Foul Play	308
Record Breaking	311
Coming Up	326
About the Author	345

INTRODUCTION

SOMETHING IS WRONG.

A pair of headlights zoom into my F-250 tow mirrors. The car charges like a bat out of hell and sling-shots around my pickup truck. The man inside that car is not driving like a normal human. With a growing sense of dread, I look through my driver's window. The car moves away, the way you'd pull your fist back in a bar-fight. For a moment he lingers along the freeway center divider, with his tires slinging dirt and debris from the emergency lane.

But he doesn't stay there. He slams his steering wheel and pile drives towards us like a wrecking ball. My whole body tightens. This will hurt. The car smashes into my front quarter. There's an explosion of shiny plastic and glass as the car hits the worst possible place, my front steering wheel. This is no accident. This is murder.

Police call it a "PIT" maneuver. It causes a driver, in this case me, to lose control of the vehicle. But there are no police here, just a hoodlum, hell-bent on killing.

Surprisingly, my truck is still on the road. I take stock. The truck's not wobbling. All parts seem to be working. But now what can I do? Normally, you pull over and swap insurance info after an accident. But I can't stop in the freeway with this guy trying to crush me into a ball. Stopping would make things worse. All I can do is hold onto the big steering wheel of my old truck as the man winds up for another smash. There are no cars in my rear-view mirror — no cars in front. He waited until no one was around. I glance at the child safety seat next to me.

"Did you feel that?" I ask.

My tiny son looks up, unaware. I think he can sense from my tone of voice that we're in trouble.

BANG! The man rams his car into our truck again. Plastic parts fly over our heads as the killer leans on his steering wheel, trying to push us upside down. But the old Ford doesn't give an inch.

They don't build trucks like this anymore. New trucks are lighter, smoother, and quieter. But the sturdy old turbo-diesel has the frame and engine of a grain harvester. Thank God our old Ford F250 doesn't even shudder as the attacker rams all the way around us with car parts and glass exploding in air.

"What is it?" Wyatt asks. Our truck is so high that my five-year-old son can't see Mad Max outside.

"That car just hit us, buddy," I say. "Hold on!"

The man's car is smashed, and he's coming back to hit us again, in a Moby-Dick-style torpedo run. Thank God Wyatt and I are wrapped in eight thousand pounds of super-duty steel as we careen down I-210 with a crazy man trying to kill us.

I grab my phone and dial as fast as my fingers can go.

"911, what is your emergency?" asks the operator.

I try to sound calm, but that's impossible.

"A guy just hit us with his car!" I yell.

"Where are you?" asks the operator.

"Whoa! Hang-on! He's going to hit us again!"

I'm glad the 911 lady is on the phone, but I can't give her an address when this guy is crashing into us.

I grip the steering wheel and brace for impact.

"Watch out, Dad!" yells Wyatt, who's now peering over the dash to see the assassin.

"Why is he doing that, Daddy?"

"I don't know buddy. Sometimes people get angry when they drive."

It's true. Los Angeles is the road rage capital of the world. But this isn't road rage. The man was nowhere near us before he attacked.

Maybe I should have listened when I was warned not to talk

about the crime at the school bus stop. Everyone else kept quiet. Even the media are afraid to report, for fear of reprisals.

Uh oh! Here he comes again. The man drives a hungry circle around us the way a shark circles prey before coming in for the kill. Fortunately, hitting a super-duty Ford with a passenger car is like hitting a bull-dozer with a banana. The man's sedan is crumpled, and Wyatt hasn't moved in his child safety seat.

Suddenly, the killer rockets in front of us. As he zooms away, I take a breath and get back to the 911 operator.

"We're on I-210 and Lincoln," I say. "There's a crazy man hitting us with his — uh oh — "

Smoke pours from the killer's tires. Having failed in his first attack, he's now trying to hit us from the front.

I slam on the brakes to avoid running into him, and I prepare to grab the Smith and Wesson that's always under my shirt, ever since I started fighting against organized crime.

Poor Wyatt. He's just five years old. He shouldn't be here. He's seen and heard a thousand things no toddler should. Ever since I told gangsters to leave the school bus stop, Wyatt's childhood has been car-jacked by killers.

I hate that my little one is exposed to evil. But I had to make a hard choice, and I chose to make the town safe. That means getting the organized crime ring away from kids at the bus stop.

And now we're caught in a life-and-death struggle. I look left and right of the killer's car, but there's no escape. All I can do is push harder on the brake pedal. Underneath us, a Ford micro-computer makes a hundred calculations in a second to carefully grip and release each brake caliper to keep the old truck from sliding out of control. We lean left, then right, front, then back. It's like riding an aircraft carrier in a full-throttle evasive maneuver.

A minute ago, eight thousand pounds of steel kept us safe. But since the killer couldn't smash us, he's using our weight against us. And I can't see a way to stop the truck before we hit.

As smoke pours from the assassin's tires, I see movement inside the car. Is he planning to jump out and shoot us? I'm trying to get us out of harm's way. But stopping this truck is like stopping a freight train. I rise in my seat, standing on the brake pedal, but my heart sinks as we careen toward the killer. I don't have enough room to stop this truck. We're going to hit.

Wyatt and I slide forward in our seat-belt straps. Gravity has us.

I've faced death a thousand times. If you read my first book, you know I have a gift for getting into trouble with no escape. But this time is different. This time, I brought my little boy with me.

I shoot my right arm across Wyatt's chest to brace his tiny body for collision.

Shit! How many bullets does my gun have in it?

Welcome aboard my ill-planned journey. Please fasten your seatbelt and brace for impact. I can't promise you won't be crushed to death or shot by gangsters. But I promise you that I won't quit until the town is safe again.

My name is David Alan Arnold. And this is my story.

THE HARD WAY

I SHOULD HAVE TAKEN THE EASY way, like everyone else. Standing up to gangsters isn't smart. But a good friend once told me that I'm, "thick as a brick." So here we are, sliding toward a killer on Interstate 210, with the 911 lady on the phone.

Although I don't have room to stop my eight-thousand-pound truck, we magically haven't hit the killer's car yet. I rise in my seat, pressing my foot against the brake pedal.

Suddenly, the killer pauses. I sense fear and frustration. He's tried multiple times to get me to lose control, but Wyatt and I are still on the road. And our big, ten-ply, all-terrain tires are grabbing the city asphalt with all their might. We're starting to slow down. We might not hit as hard as I thought. But the man doesn't wait to find out. He suddenly slams his wheel to the right and stomps on the gas. The smashed sedan rockets toward the Lincoln Street exit, leaving a trail of smoke and car parts.

Finally I get back to the 911 operator. "OK. He just exited at Lincoln," I say.

"I'm updating officers," she says while typing. "Is your truck still drivable?"

"Yeah. I think it's OK."

I check my gauges. All are green. I have no red warning lights. The old truck sounds OK. There are no vibrations.

"I put out a bulletin for officers to be on the lookout for that car," she says. "You can come in tomorrow and fill out a report."

That's a good idea. It's almost bedtime for Wyatt. I shouldn't keep him up late fighting crime. He didn't ask for this. But you can't choose your family. And my little one got the only dad on our street who didn't back down when killers took over our town.

You might be wondering how gangsters can take over a town in modern America, in full view of police and news media. How can thugs be allowed to deal drugs and death at a school bus stop?

Well, some members of Law Enforcement say there's no problem. After repainting several buildings, changing documents, and removing some bodies, they say nothing bad or illegal happened. But I'll let you decide. I'm going to show you everything I saw at a hotel and medical facility that authorities say do not exist.

To begin, I'll have to take you back to a terrible time in my life. First, the love of my life, AKA Wyatt's mom, left and broke up our home. Then my work partner was killed in a helicopter crash...

QUITTING TIME

THE AVERAGE PERSON SPENDS A third of their life sitting at a desk. But in my office, we fly. My pilot pulls his collective lever to cruise power, and we rocket through a desert landscape. The helicopter shudders as we careen a few feet above rocks and Joshua trees. As we roll left and right, our windows flash back and forth with a blurred picture of Planet Earth and a crystal blue sky above.

But in spite of the thunderous noise, the bouncing, shifting, and rolling of the helicopter, the picture in my Cineflex monitor sits perfectly still. The camera is gyrostabilized by advanced aerospace circuitry, mechanical isolation, and a motorized gimbal. The Cineflex picture is completely smooth, as if an angel is carrying the camera inches above the earth.

My pilot pulls back on his cyclic stick and lowers the collective. We slow to a hover.

"This is the place," he says. His normally cheerful voice is heavy with sadness.

I look down, below our skids, to a scarred patch of earth.

"He hit here and ended up over there." My pilot points to the place where our friend and work partner, David Gibbs, came to rest. One minute he was a sky king, flying for a network TV show. The next minute, he was gone.

Looking into the lifeless desert landscape where my friend lost his life reminds me of the day I heard the news of the crash. My life has never been the same. I wish I'd never heard that news. I wish Gibbs was still here to give me shit with every waking breath. But this is where Gibbs "entered the sky" for the final time. The laws of physics passed judgement, and I never saw him again.

"Damn," I say. "I sure do miss that guy."

My pilot pushes forward on the cyclic, and we continue on in silence. In half an hour, he pulls back and hovers again.

"This is it," he says.

They hit here and ended up over there. My pilot points to the resting place of another one of our colleagues, who'd crashed and died while flying for a TV commercial.

"I'm retiring," says my pilot.

My eyes grow wide. I stare at him, speechless, for a beat.

"What?"

"I've been doing this too long. I've flown over too many of these accidents. I'm gonna hang it up."

The shock washes over me and gives me a chill. I've been flying with him for over a decade. I'd never thought of him stopping, ever. But I can tell from his tone that his mind is made up. I guess it makes sense. I know he's right. No one lives forever, especially in my line of work.

"WE CAN'T TELL YOU. IT'S A SECRET…"

Back at home, I sit in an empty house.

These are dark times.

Wyatt's mom split.

My friend and work partner, David Gibbs, was killed in a helicopter crash.

I'm not working as much as I used to.

I'm running out of money.

Bills are piling up.

I always save money for hard times, but the judge took my savings account and gave it to Wyatt's mom and her lawyers.

Fucking lawyers.

Without work, I have to borrow money on credit cards, and then I have to pay thousands of dollars to mom and her lawyers each month.

Little Wyatt could spend his days with me on Cloud Nine. But instead, his mom locks him up with a nanny and sends me a bill for childcare. Instead of time with Wyatt, I get another bill I can't afford to pay.

Normally, I'd be getting a call from my work partner, David Gibbs, and we would fly on an adventure. But since his accident, the phone doesn't ring like it used to. I found out some people believe I'm dead. Gibbs and I flew together so often that, when people learn of his crash, they automatically assume I crashed with him. So they're not calling to hire me. I guess it makes sense, but this is awful timing. My PO Box is full of unpaid bills.

But things could be worse. At least I can spend some quality time at Cloud Nine — my happy place on Earth. I take a deep breath of cool mountain air and watch every color bird fly up to the giant oak tree in front of the house. The trees form a green cathedral around me. The sound of birds and breeze is like a lullaby for my spirt. But one of my neighbors walks over with a pained expression. He looks like someone has died and says, "Dave, there's something going on in town. There are scary-looking men in vans."

"Scary-looking men? In Sky Forest?"

So I step off of my porch and go for a walk. As I round the corner, I see something at the school bus stop. My neighbor was right. Scary-looking thugs pile out of a rusting cargo van. The group gathers the way convicts do in a prison yard. They shoot murderous looks at me and puff on cigarettes. Unlike the peaceful townsfolk of Sky Forest, these new arrivals look like drug dealers. Tattoos of knives and venomous snakes poke out of baggy-clothes. Cigarette smoke billows around sideways ball caps.

"Whatchu lookin at, motherfucker?" says a tattooed thug. He grits metal-encrusted teeth. Homemade prison tattoos cover a leathery hand that reaches toward his pirate goatee to grab a smoldering cigarette.

What the hell is this?

Van after van, doors swing open, and ruffians pile into what used to be a chiropractor's office. But where is the chiropractor? What happened to the kind doctor who used to let the kids at the school bus stop use his parking lot for shelter in snow storms? Now the doctor's parking lot is full of gangsters.

It looks like the chiropractor has been replaced by Murder, Inc. But there were no signs. There was no public hearing for a change of use. How can an army of tattooed thugs take over a medical office?

Whatever the crazy business is, there isn't even a sign saying the name of the new company, just hundreds of people moving back and forth as if on a hell-bent mission. I flag down the nearest thug. "Excuse me. What kind of business is this?"

A frown forms over his ZZ-Top beard. The man's body goes rigid. He postures like a feral cat. His jaw tightens, and he seethes, "We can't tell you. It's a secret."

One of my neighbors snaps a picture, and a tall gangster torpedoes out of the building and grabs for her camera. Scary-looking men shout profanity and threats through cracked van windows.

Another neighbor tries to snap a picture of the gang with her smart phone, but a horde of thugs dog-pile out of the building and surround her, taking pictures of her with their smart phones. My terrified little neighbor flees to her house. She won't do that again…

"Look at this cowboy! Go home, asshole!" I hear from inside a gangster van. Normally, I don't mind people making fun of my hat. But why are these guys acting so nasty at a school bus stop? I guess foul language and combat posturing are normal where they come from, but this is where children gather six times a day to wait for a bus. So I don't move a muscle. I smile but look each of them directly in the eye. I don't know what this is, but I don't feel like moving or blinking.

A quick head-count tells me the gangsters outnumber the residents of our little town. Maybe that's why they feel emboldened to shout profanity across the street. The scene looks like a Wild West movie, where a criminal gang takes over a little town with no sheriff to stop the bad guys.

Since Gibbs' death, I'm not getting calls for work, so I have plenty of time to keep an eye on the neighborhood. I set my watch and return to the school bus stop every time the kids are waiting for a bus. I snap a picture of the kids standing next to the cargo vans. This causes a riot at the "secret" business.

I can hear shouting from inside the building. A worker hurries out of the school bus stop and walks across the highway to where I'm standing.

"Stop taking pictures," he steams.

I say, "Excuse me? Do you see where you're standing?" I point to the school bus stop. "My name is Dave Arnold. Who are you?"

"We're here to stay, Dave."

That's weird. I didn't ask him how long he plans to stay. Is he aware that he shouldn't be here and is using reverse psychology? Again he demands that I stop taking pictures.

"What are you doing at the school bus stop that can't be photographed?" I ask.

With death, financial ruin, and the loss of my family life, the one good thing I had was the peace and quiet of my home in Sky Forest. But now that's been ruined by a horde of ruffians.

I'm not in the mood for this. I let the rude stranger know I'll be taking pictures and keeping an eye on the kids.

SKY FALLING
A Documentary

He huffs, "This conversation is over."

But I don't think the conversation is over. That conversation was just getting good. So I stand right on my spot and stare at him and his "secret business," and I snap more pictures.

Seeing I'm still here, he comes back across the highway to where I'm standing. He's much more polite this time. He introduces himself as Kyle Avarell and hands me a business card. It says "Above it All Treatment, Operations Manager."

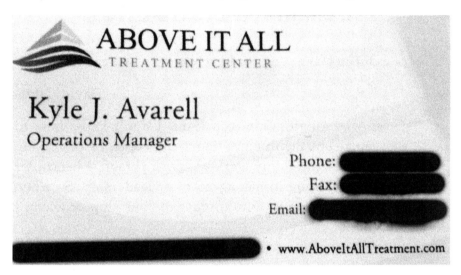

I ask him what "Above it All" is. He says it is a drug rehab, treating drug addicts at the school bus stop. I think carefully about what he's saying. Kyle has more drug addicts with him than people who live in this town.

His gang business doesn't look legal, and I know it's not safe around school kids. I don't know what to do, but I have to do something. So I say, "I want to speak to the owner of Above it All."

"No," he says. "You can talk to me. We're here to stay, Dave."

As Kyle walks back to the school bus stop, I take a deep breath. My throat tightens. What kind of gangsters set up shop at a school bus stop?

I don't know why this is happening. But a thought comes clear in my mind. I have to do something.

WEEKLY MEETINGS (HOW I MET MY NEIGHBORS)

"You're a dufus Dave. You could screw up a two-car funeral."

— David Gibbs

GIBBS WAS JOKING. BUT HE'S right. I don't know how to organize stuff with more than two people. I work on big TV shows, but I only work with one person, a helicopter pilot. I have no training in management or leadership. So when I invite the entire community to discuss the bus stop problem, I know it might not go well.

A motley crew of townsfolk assemble. The first thing I notice is that my neighbors don't get along. In fact, getting them to agree on anything is impossible. Talking to this group is like trying to herd kittens. Kittens are cute, but they only want to wander in their own direction. Any suggestion of a direction causes them to hiss and show their teeth. There is verbal scratching and biting in the room.

Oh, well. The bus stop isn't going to fix itself. So I call the meeting to order. But before we can get down to business, some of my neighbors interrupt to say they don't like each other. Great.

But my cranky neighbors agree to set aside differences, because everyone feels the kids are in danger at the school bus stop.

A woman stands up and says she has sympathy for drug addicts. My neighbors listen carefully. I know some neighbors have suffered addiction or lost a loved one to drug addiction. There are recovering addicts in the room who haven't forgotten the struggle to win back their lives.

Some at the meeting have lost family members to addiction. I am one of them. Both of my grandfathers were alcoholics. I never met either of my grandfathers because of their drinking problems. We all share kindness and compassion for addicts and concern for people seeking treatment. But we notice that the rehab workers don't show compassion or concern. These "rehab workers" act more like street gang members than healthcare workers. People say they've witnessed drug-dealing and violence at the bus stop. Tattooed thugs say, "Don't fuck with us!" to neighbors and passersby.

"Don't fuck with us!" — Is that the spirit of recovery? I don't think so.

I've been to AA meetings and I didn't see the words "don't fuck with us" anywhere in the Twelve Steps. At recovery groups, I see humility and willingness to help each other. Breaking the law and menacing children at a bus stop are *not* the steps to recovery. This gang is bad news.

Some neighbors have experienced Above it All Treatment in other towns. One lady says the rehab was thrown out of her office building for graffiti, foul language, fighting, sexual misconduct, burglaries, and violent crimes.

"Drug addicts wandered into my office," she says. "They would blow cigarette smoke in my face and yell, 'Fuck you, bitch!' I had to call the police multiple times."

One of the rehab patients was arrested for a violent felony. He told cops he got high at the bus stop in Sky Forest—God help us.

When I talk to parents at the bus stop, they're horrified because the gang is running a drug rehab in the presence of school children. One man says, "Those guys tried to hit my daughter with their van!" He had to dive in and pull his daughter to safety as a gangster accelerated toward her with his cargo van.

County representative Louis Murray comes to one of our meetings. What he says brings shock and disbelief. Louis says Above it All is not a drug rehab. His nonsensical claim goes over like a lead balloon. One woman spits her drink.

The business Louis claims is not a drug rehab has Drug Rehab ads all over the country. A neighbor shouts at Louis, "Above it All has a drug rehab billboard on Interstate 10!" And Kyle's business card refers to the school bus stop as a drug rehab. So what's Louis talking about?

I've seen scams and rip-offs, but this one takes the cake. In spite of Louis' claim of no rehab, Above it All Treatment is offering "Naturally Effective Recovery" in nationwide ads. Gangsters are walking around the school bus stop gobbling fistfuls of narcotics. Then they stare at us the way a lion looks at prey.

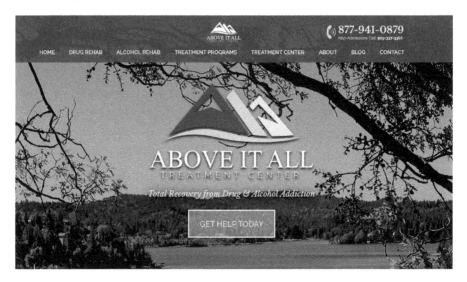

Naturally Effective Recovery at
Above It All Treatment Center

Nationwide Ad: "Above It All Treatment Center has held successful rehabilitation treatment programs for those who have suffered from addiction to drugs, alcohol, and mental disorders for many years." — At a school bus stop? You've got to be kidding me...

So why does Louis Murray say it's not a drug rehab? It's a jaw-dropping, Twilight Zone fiction. My neighbors fly into an angry fit over Louis' obvious lie. Someone yells, "If it's not a drug rehab, what is it, Louis!?"

Louis repeats the same lie, "It is not a drug rehab."

One neighbor mocks him, "That depends on the definition of 'is.'"

"He's lying!" seethes another...

Louis has angered the cats. They've forgotten they don't like each other, because now they all hate Louis.

Above it All Treatment:

"We are less than two hours from Los Angeles in a small mountain resort town...gorgeous blue skies, clean mountain air..."

The street gang is using our clean environment to make money, illegally.

Suddenly, my neighbors who can't agree on anything look like they're going to unanimously lynch Louis. I'm angry too. I know Louis is lying. I say, "Louis, we never had a problem when the chiropractor was in that building. The chiropractor never menaced the children. But now people are afraid to go near the bus stop."

Sky Falling
A Documentary

We're all afraid the thugs are going to kill someone or hurt one of the kids.

I hope the addicts are nice people. But the tattooed killers I ran into are not nice people. They should not be allowed near kids. I believe some of the bad guys still have cocaine, meth, and heroin flowing through their tattooed veins.

"You know, most violent crimes happen under the influence of alcohol and drugs," says a police officer I ran into. That may be the case at a bar or nightclub, or on the mean streets of LA.

But what the hell are these guys doing at a school bus stop? I feel like something bad is about to happen any second, with the kids caught in the middle.

One neighbor shouts, "I've seen drug deals and fights when the kids are there." My neighbor is right. Bad things are already happening.

Why is Louis covering up for criminal activity at a school bus stop?

I take careful notes during the meeting with Louis Murray. The more Louis talks, the angrier people get. Louis seems like a nice guy. But I'm shocked by his claim that the drug rehab is not a drug rehab. What kind of public servant would do this to a quiet little neighborhood?

My friend Hugh Campbell shakes his head as if he's watching someone derail a passenger train. He takes a deep breath and says, "San Bernardino County is the most corrupt county in the United States." After the meeting with Louis, I'm starting to see it.

Sky Falling
A Documentary

Louis and I share emails and notes about the meeting. Louis goes carefully, line-by-line, through all the questions we asked. But even Louis doesn't seem to like his own answers.

Louis sounds like he's a little bit nauseated when he repeats his claim that Above it All treatment isn't a drug rehab. Meanwhile, drug addicts are coming from all over the country to smoke cigarettes at the school bus stop. Some addicts keep to themselves. I hope they're nice people just trying to get sober. But some cross the street and attack townsfolk. One of them tried to run over a little girl who stepped off the elementary school bus.

I check my watch. It's time for another school bus. I ride my bicycle around the corner from my house to the school bus stop. Immediately, a small riot ensues among the "don't fuck with us" gang. "Hey! What the fuck are you doing?" a tattooed thug shouts. "Get the fuck out of here! What's a cowboy doing on a bicycle?" The thugs chuckle amongst themselves for a minute. But they're not all laughing. Some of them are looking for a fight. The gangsters hurl insults about my hat and my clothes and bicycle. Why are they so enraged?

The gangsters yell at me to go away. But I can't leave a bunch of children to a group of gangsters. So we're stuck here in a daily stand-off. The gangsters must be doing illegal things at their secret business. Otherwise, there's no reason to care if someone is standing across the street from them.

But I don't know what else to do, except watch over the kids and make sure no harm comes to them. So I take a deep breath, smile, tip my hat, and say, "Good morning guys. Do you know where you are?" When I make eye contact with the bad guys, some of them look away. And some of them look at me the way a lion watches people through the cage at a zoo — like a piece of meat. But I don't blink. I'm not moving. There are elementary school children standing a few feet away from these animals. One of them swallows a fist-full of narcotics and shoves the guy next to him. "Don't touch me, bitch! I'll fuck you up!" I can hear him from across the highway. But the kids are standing right next to him. God help us.

Maybe the media will help. I contact investigative journalists at TV stations and newspapers. But not one of them responds. That's a shame. No one outside Sky Forest knows what's happening here.

Back at our weekly meetings, my neighbors lean in to hear what they missed and share what they saw during the past week. "OK guys, let's get together every week to compare notes and share information." I say. By a show of hands, we assign volunteers to investigate different aspects of the "don't fuck with us" gang. My neighbors are not criminal investigators, but everyone pitches in to gain insights on the business.

When I call County zoning officials, they say, "That drug rehab is illegal. Under no circumstances can you have a drug rehab at that address." So that's it. This is an illegal business, and certain county officials appear to be allowing it. That makes me nervous. If corrupt officials won't protect the kids from an illegal business, we're in big trouble. Bad guys are breaking the law. But so are the authorities.

We're all in danger. The bad guys already tried to run over one of the kids with their serial-killer van. But authorities refuse to help. So I slide .38 caliber bullets into a revolver and holster it under my shirt. It feels surreal to have to carry a lethal weapon in what used to be one of the safest towns in America.

MONDAY NIGHT FOOTBALL

BACK AT WORK: The theme song for Monday Night Football echoes through my right ear. My left ear rattles with the non-stop chatter of Philadelphia Air Traffic Controllers. I glance down at the on-air monitor to see a cavalcade of computer animation dance across the screen. As Hank Williams Jr. music blasts to a crescendo in my Motorola earpiece, ten million people lean in to their TVs to see my shot of the Philadelphia skyline and Lincoln Financial Field. But just as Hank hits a high-note, a new sound hits my left ear, "Thirteen's no good!" shouts a TV producer.

Shit! I'm Camera Thirteen. A third radio rattles with another voice, "Dave? Talk to me, buddy. Do you see that flickering on your screen?" Tony Vasquez, our senior wireless engineer, calmly troubleshoots from underneath a fifty-three-foot tractor trailer, where a thousand wires carry high-def pictures from live cameras around the football stadium into the A-Unit, where our director commands a hundred people to pan cameras, blast lights, and cue graphics to start the show for Monday Night Football.

Tony is calm, but his voice is energized, because of what we both know—if our camera isn't ready in exactly four seconds, the director, producer, team-owners, and 10 million fans will see static instead of a glittering stadium crowned with fireworks.

I glance at my Cineflex controls. Everything looks perfect. Orange lights blink as Observer One bounces against the clear-air turbulence. Outside the airplane, our Cineflex camera stabilizer twitches with every wind gust. The plane bounces left. The turret bounces right. The plane bounces right. The turret bounces left.

The result is a perfectly smooth picture of the stadium, even though I'm being thrown like a rag doll against my seatbelt straps.

WHAM! Wind shear hits the plane, pushing us so violently that I hit my head on the ceiling. "Uhhhn!" Captain Dan groans against the g-force that kicks us up twenty feet, then down thirty feet. "Hang on, Dave!" Dan shouts through the aircraft intercom. And that's when I see it—a tiny flicker in my monitor screen.

My left hand shoots to a walkie hand-mic. "I see it, Tony! It's on my end" I say. Tony breathes a sigh of relief under the TV truck. Neither Tony nor I know why the picture is flickering, but at least Tony knows the problem isn't in his equipment that feeds our live picture to TV screens around the world.

"OK buddy," Tony says. "We're live in three seconds. Figure it out."

"Thirteen is no good!" a producer yells to warn the director that our opening shot cannot open the show.

"Dave, are you good?! Can I use you?!" the director pleads. It would be a huge embarrassment on national television if the first thing 10 million people see is my messed-up picture. Normally I would grab my third mic and answer the director, but there's no time to talk. I have exactly three seconds to fix this.

But I don't know why the picture is flickering. The camera was perfect until we pushed through this turbulence to open the show. Ah. The turbulence. Something must have gotten knocked loose in the gimbal that stabilizes the camera.

"Three, two, one!" the director counts down to my live shot. But my camera is still flickering. "Hunnghh!" I grunt as another wind gust hits the plane. BANG! My head hits the ceiling again. The world around me is a blurry picture of wind and vibration from Observer One's twin props pulling at the tumultuous air over the brightly-lit stadium. Now I have half a second to fix this camera. No time to think—I let go of the Cineflex joystick and grab the gimbal control cables. I squeeze with all my might to grip both cables in one hand. While squeezing with my right hand, I grab the Cineflex joystick with my left. I'm not left-handed, but I'll have to make this work.

"Dissolve thirteen!" the director shouts. I squeeze the cables as the tech director reaches across a hundred brightly lit buttons in the mobile TV studio to find the one marked "13."

"Thirteen! You're on!" shouts the director. I look down at my monitor, and I want to kiss it. The picture is good. I glance at my on-air monitor and see Monday Night Football graphics fly across our perfect picture. "Thirteen is good!!!" shouts a producer in the TV truck.

"And zoom Thirteen!" shouts the director. "Bring us home Dave!"

I know what the director wants, I've done it a thousand times. I'm supposed to tilt the camera with my right hand on the Cineflex stick and push the zoom lever with my left hand to fly the camera from the city into the stadium, right down to the bright green grass and football players, who are getting ready to kick off. This is how we start every Monday Night Show. But I can't steer with my right hand. It's stretched like a clam-shell across the two cables. I can't zoom with my left hand. It's holding the joystick All I can do is keep the camera steady.

"I got it Dave!" Captain Dan shouts into his Bose Headset. With the experience of a hundred Monday Night Football flights, Captain Dan throttles up our twin Lycoming engines. For a moment my ears fill with the ear-splitting thunder of eight boxer cylinders as Dan pushes his throttles full-forward. I'm pushed back in my seat as we accelerate, and Dan zooms our entire airplane toward the stadium. It's a perfect move. Fireworks fill my frame as the Eagles kick off.

"That's great, Dave!" shouts the director. "Ready Camera Two. And dissolve Camera Two."

"We're clear!" I yell at Dan. We did it. As usual, Tony, Captain Dan, the producers, and the Cineflex did just the right thing at just the right time to get us on the air in front of millions of sports fans. Now I just need to un-pretzel myself so I can fix the cables and put both hands where they belong for the next four hours of football.

After the show, Captain Dan lands the plane so I can make my way to an airline counter, and then Cloud Nine. I extend my hand to Captain Dan. "Thanks for the safe ride."

And with that, I turn away from the fireworks and roaring crowd of Monday Night Football and go back to my quiet life in Sky Forest. I take a deep breath. I used to look forward to this. I used to get excited as a kid at Christmas to come home to my quiet mountain home. But the beauty of the forest has been taken over by a gang of thugs. The thought of going home isn't the same.

SIGNS

As I walk past Philadelphia shops, I see posters and flyers for the Eagles. Fans are desperately hoping to make it to the Super Bowl. That's when I see it— a piece of paper taped to a store window. It says, "Change of Use" and "Public Hearing." Apparently there's a process if you want to change the store into something else. That reminds me of how the school bus stop in my little town changed into a drug rehab.

But regulators already told me the rehab is running illegally. I wonder how the illegal rehab got permission to use the school bus stop in Sky Forest without a "Change of Use" or a "Public Hearing." So when I get back to Sky Forest, I ask business owners what they had to do to get their businesses open in Sky Forest. I'm shocked by their answers.

"The county wouldn't let us open," says a local shop owner, who runs a tiny shop that doesn't endanger anyone.

"What?"

"They held us up for two years with parking permits, health codes, building, and safety. They took all our money."

I can't believe what I'm hearing from a sweet lady who runs a renowned business in Sky Forest. She says, "After two years of red tape and hundreds of thousands of dollars in licenses and fees, the county still won't let us open. We told the county, 'We're broke. We quit.'"

As we're talking, two ladies walk into the very same shop with their purses out, ready to make a purchase. The shop is open for business. So I turn back to the business owner and ask, "If the county wouldn't let you open, then how did you do it?"

"Janice Rutherford."

"Who?"

"Our county supervisor," she says in a whisper, as if someone will hear and punish her.

"Why are you whispering?" I ask.

"Dave, you can't tell anyone about this. They'll close me down."

I am gob-smacked. County regulators run the government like a Godfather crime family. And this sweet shop-owner is worried about retribution.

I guess that's why law-abiding business owners had to spend years and hundreds of thousands of dollars to get permission to open. Meanwhile, criminals are running an illegal business at the school bus stop without a single license or permit. But what does a county supervisor have to do with that? Aren't there bureaucracies full of inspectors and regulators who govern these things?

When I call the bureaucracies and regulators, I'm told that there was no "change of use" hearing to turn the school bus stop into a drug rehab. In fact, there are no records at all. It's like the business doesn't exist—just like the Mafia.

A county worker says, "That drug rehab is illegal. They don't even have an occupancy permit."

That's surprising, because the muffin lady two doors down from the bus stop was shut down and boarded up by county regulators, but a dangerous business with criminal activity is allowed to run with impunity. How can it be so easy for criminals and so hard for the muffin lady?

Hugh Campbell is right. San Bernardino County is corrupt. And the lions at the school bus stop look hungry. I'm afraid someone is going to get hunted down and killed.

I head down to the Building and Safety Department to check the zoning law for the school bus stop. The zoning law says, "Medical Rehab: Use Not Permitted." Next I take a look at the drug rehab industry. I find that the school bus stop rehab is violating the ethical principles of recovery by sending drug addicts to smoke cigarettes next to the kids.

Anonymity is key to recovery. If an accountant tries to get off pain meds and one of the kids recognizes him at the bus stop, the accountant could lose his job. Would you hire an accountant on drugs? If a kid or parent talks about who they saw in treatment at the school bus stop, an employer or competitor could fire the addict or hurt their business. If an addict loses their job because they're seen at the school bus stop, they can be driven back to drugs by the stress of losing their job or their home.

This is life and death stuff. Identities of addicts must be protected. But the gangsters didn't even put up a privacy fence. What kind of rehab doesn't protect their patients?

I call the rehab myself. The number was easy to find in their nationwide ads that pop up every time I open my computer. If you have internet access, you've probably seen an ad for this drug rehab. I dial the number, and the lady on the phone says it costs $135,000, but she doesn't mention that I'll be crammed into an illegal facility at a school bus stop. This is a scam. And it's being sold, nationwide, to drug addicts and health insurance companies.

It's a con-artist's dream. People will pay $135,000 for help with a problem that's ruining their life, and they're right to assume that medical ethics and regulations are protecting identity and safety. But that is not the case. Victims don't realize they're walking into a scam run by a street gang, and the authorities are looking the other way.

"Dave, do you own a gun? Do you have a gun in the house? Those criminals are making a lot of money..." A local shopkeeper worries about my safety. She would know. She went to hell and back just to get her store open, and it cost her dearly. Crooked officials are threatening to close her down as we speak.

The shopkeeper is right. You can't fight city hall. So, I slide twelve-gauge cannon shells into a pump-action Winchester. I don't want to keep a loaded gun where my toddler sleeps, but someone has to protect him. The authorities aren't doing their job in Sky Forest.

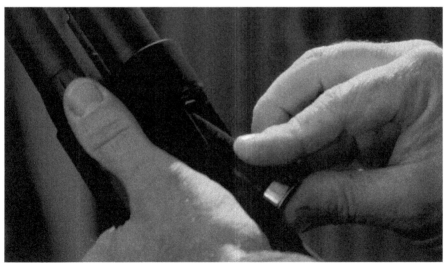

GARBAGE

I GRAB WYATT'S HAND AND WE walk out of our house to find a familiar sight. Our neighbor's trash is spread across the street. "C'mon buddy," I say. We walk over, and Wyatt watches as I pick up half-eaten fast food packages and throw them back into my neighbor's garbage can, which has bear tooth and claw marks.

"Watch out Wyatt. Don't step there!" I yell. Wyatt looks down. His little shoe is covered in slime. It's hard for my little one to know where to put his feet. The bears dragged half-eaten garbage all over the place.

This slimy, stinky scene repeats every week, because crooked county officials force every homeowner in Sky Forest to pay for this trash service. Even if no one lives in the house, you'll get a bill. I get a bill for trash collection even though I take my garbage to the landfill to keep it away from the bears.

I don't want the bears to replace their normal, healthy diet with my garbage. So I don't use the county's trash service, which requires me to leave trash cans where bears can rip them open and eat the trash. But even though I don't use their trash service, I still have to pay for it. If I don't pay for the trash service, the county will put a lien on my house.

"Watch out buddy!" I yell at Wyatt. Wyatt jumps when I point down at his ankles, where my neighbor has placed shards of sheet metal for the trash guys to pick up with her trash. Wyatt could have cut his little legs on the sharp edges. But what about the bears? They come to my neighbors' trash cans every week because they know the cans will be on the street, full of junk food garbage. But the bears can be cut by that sheet metal just as easily as my son.

When crooked politicians make a shady deal to squeeze more money out of homeowners, it's politics as usual, I suppose. But this is dangerous.

STORYBOOK

WE LEAVE THE TRASH BEHIND and walk around the corner to St. Richard's Church. As we walk up the driveway to the church, a commotion catches my eye. There's something going on at the hotel.

"What's going on at the Storybook Inn?" I ask a neighbor who's getting out of his car. My neighbor grimaces as if in pain. "The rehab is taking it over," he says.

"You've got to be kidding me. They're taking over half the buildings in town. There's no way this is legal," I say.

"There's nothing legal about those guys." He gestures to the gangsters across the street but seems careful not to make eye contact with them.

I do a head-count across the street. My neighbor is right. A half-dozen guys are posturing like a street gang. One of them puffs out his chest. I guess I looked at him for too long. He starts to walk toward us, a heavy wallet chain swinging like a wrecking ball above his biker boots.

I grab Wyatt's hand and lead him inside the church. Normally I take time to talk with the guys at the rehab, but not when Wyatt is with me. We gotta go before the tattooed gang-banger tries to start trouble.

The Episcopal Church service is liturgical. The Vicar reads, and the congregation reads back. But there's an open talk at the end, and the Vicar kindly asks if there are any prayer requests. I raise my hand.

"Can we pray for the addicts at the hotel across the street?"

The Vicar nods and says a beautiful prayer for the patients and workers at the hotel.

As the service ends, I grab Wyatt's hand and walk him out of the church. I'm careful to keep my little one on the opposite side of the street from the gang that now occupies the hotel. The traffic is closer to us, but the cars don't have arrest records like the men who are mean-mugging me in front of the hotel.

The Storybook Inn used to be a charming bed and breakfast. But that changed over-night, without a public notice or a hearing. I already know they don't have a rehab license. Maybe that's why they kept the hotel sign up. That might be confusing to the traveling public. But I don't believe these thugs have a rehab license. So I suppose keeping the hotel sign up will keep people from asking too many questions about their illegal business.

On the other hand, I fear for any tourist who sees the hotel sign and wanders in looking for a hotel room. "Lodging and Dining" sound nice, but I know the truth. The hotel staff has been replaced by a gang of thugs.

Our peaceful little town suddenly looks like half street gang, half zombie apocalypse. Hordes of intoxicated addicts wander the streets day and night. In the wee hours of the morning, they walk across our yards, partying and yelling. If you challenge them, they might have a nice conversation with you, or they might come after you with a knife.

Crash! My neighbor hears a noise in front of her house. She fearfully climbs out of bed and peers through the curtains. A half dozen hooded figures walk through her yard. One of them stops. He leans against her oak tree and vomits outside her window.

Drugs do mysterious things to the human body. Because there are no security barriers at the school bus stop, we have a lot of unexpected encounters with drug addicts in various states of sobriety. Sometimes they quietly walk by. Sometimes they break into our homes.

Another neighbor clicks the remote on his keychain. His car lights flash and he walks toward his front door like he does every night, but tonight, something is different. There are shadowy figures sitting in his patio chairs. The hooded men are talking loudly and appear to be drinking alcohol. My neighbor slips out of sight and goes in a side door to get his gun. When he points his gun at the drunken hoodies, he discovers that the men came from Storybook Inn. The liquor they are drinking is his liquor, taken from inside his house. The hoody mob broke in and stole his liquor. Now they're sitting in his chairs getting drunk. He frog-marches them back to the hotel. And so ends a typical night with our new neighbors at the Storybook Inn. Thank God no one got stabbed or shot.

There are now more drug addicts than residents in Sky Forest. Maybe that's why there were no "Change of Use" or "Public Hearings." The laws don't allow this criminal chaos at a school bus stop and hotel. The scam wouldn't survive a single public hearing. But organized crime doesn't do public hearings. So here we are, overrun by gangsters who run an illegal company that doesn't exist on paper.

The "don't fuck with us" gang gives Storybook Inn a creepy vibe. This ain't Betty Ford. They don't welcome with open arms. They storm around the place and threaten anyone who looks too carefully at their hotel facade.

I suppose the bad guys believe that if the neighbors are too afraid to talk about what really goes on in the "hotel," they can keep the fraud going. The gang relies on a steady stream of intoxicated customers from all over the country who have no idea they're entering an illegal facility. After hearing authorities lie about it, I can understand why Kyle Avarell thinks the school bus rehab is "here to stay." Who will make them leave, if not the county regulators who shut down the muffin lady? Who will arrest them, if not authorities who say the business doesn't exist? If the business doesn't exist, they can't be tasked with shutting it down.

I pull a heavy set of work gloves on and pick up a soda cup and a cheese-burger wrapper someone threw out of a car window. I drop the trash in a giant black trash bag. Today, I'll pick up trash from the post office all the way to Santa's Village.

I've been doing this for years. But ever since the rehab moved in, I noticed a change in the road-side litter. As I drag my trash bag toward Storybook Inn, the burger wrappers and soda cans turn to cigarette butts and whiskey bottles. I reach down and fill my gloved hand with a fist full of those tiny liquor bottles they sell on airlines. I guess that's how you sneak alcohol into a drug rehab. The empty liquor bottles explain some of the behavior I've seen at the school bus stop. Some of the rehab patients aren't entirely sober. And some of the rehab workers talk in a strange way, as if they're under the influence, too.

As I walk past the Storybook Inn with a heaping trash bag of beer cans and whiskey bottles, the meanest men I've ever seen track me with murder in their eyes. They watch me all the way across their illegal driveway.

As I leave the creepiest hotel in America, the road-side trash turns back into normal highway debris. And that's where I find it. Santa's Village.

Santa's Village is also a tourist business, like Storybook Inn. But under the current county regime, Santa's Village isn't allowed to open for tourists. The same county regulators who allow gangbangers at the school bus stop have padlocked Santa's Village. As I look across the vast emptiness of the parking lot at

Santa's Village, I'm holding a heavy bag full of whiskey bottles from the Storybook Inn.

I'm still trying to figure this out. It seems like if I just notify the right authorities, they will shut down the dangerous, illegal rehab on the spot. So I call the county again and end up talking to Janice Rutherford's office about the rehab. They repeat the same lie, that Above it All Treatment is not a rehab. Hmmm, that's not what the rehab website says. The website at www.aboveitalltreatment.com says, "Above it All Treatment is a rehab with counselors standing by to treat drug and alcohol addiction."

When I complain to county regulators, I'm told the rehab doesn't have county approvals. So the place continues to run illegally.

A sign in front of the place invites the public to lodge and dine. I wonder how many people entering the hotel could end up being harmed by gangsters posing as hotel workers.

The killers at the Storybook Inn stalk me with their eyes, even when I'm just picking up the trash they throw onto the street. They don't act like hotel workers. They keep a lookout the way you would if you were robbing a bank or hiding a dead body.

The kids at the school bus stop are in danger. But I don't know how to fix this. Without law enforcement, the only thing I can think is to go online and warn the traveling public. So I review

Storybook Inn and Above it All Treatment on www.yelp.com. I hope it saves someone from unknowingly entering a real-life Bates Motel.

Drug addicts nationwide are lured to Storybook Inn with a promise of professional healthcare. But after they give up their wallets and cell phones, they find themselves caught in a trap. They are trapped in a hotel the county says doesn't exist. I know the hotel is secretly run by outlaws. If the outlaws murder the customers and steal their health insurance checks, would anyone even know?

So I review Storybook Inn and Above it All Treatment under hotel and drug rehab categories, even though they're at the same address—a school bus stop. That is officially the weirdest thing I've ever done on yelp.com. As you can see from the happy green star, I was "First to Review" them.

I recommend you bring a gun if you plan to stay in the hotel overnight.

David A.
Los Angeles, CA
4 friends
74 reviews

Share review

Compliment

Send message

Follow David A.

 5/28/2013

First to Review

I checked out this facility in-person and online.
There appear to be some ethical lapses.

Above it All Treatment offers intensive, drug and alcohol rehab at a school bus stop. (971 Kuffel Canyon Rd. Sky Forest, CA).
The client smoking area is in the same parking lot where parents and children gather for school bus drop-off, pick-up.
There have been media reports on this issue. See CBS 2, Greg Mills report on Sky Forest Rehab, Above it All Treatment.

I was told by an Above it All Official that, overnight housing is located in a hotel called The Storybook Inn, which is open to the public. The State of CA shows no licensed rehab at that address. San Bernardino County Officials said this facility is a Hotel, not a Rehab.
I saw van-loads of Above it All Clients loading in and out of the hotel, confirming Above it All statements.
Above it All Treatment is offering hotel accommodations in the same building. I was told by owner, CEO, Kory Averall, that I could enter premises and reserve a hotel room. But his business card and company web-site indicate Drug and Alcohol Rehab/Detox.
I'm very concerned for the ethical treatment of clients who enter here for help with addiction. I saw no protection of clients' privacy, nor any separation from the general public at Above it All Locations.
In fact, they appear to be selling services to Rehab Clients and Hotel Guests in the same building.

Buyer Beware.

Was this review …?

 Useful 39   Funny 8   Cool

THEY'RE ALL CRIMINALS

I ask Louis Murray for law enforcement to help us at the school bus stop. So at one of our weekly meetings, Sheriff Captain Rick Ells makes an appearance. He looks nice. He's polite and well-spoken.

I introduce Captain Ells to a crowd of my neighbors. I try to give the Storybook workers some credit. "With all due respect to the rehab workers, they're not all criminals—"

"No," the captain interrupts. "They're all criminals. I checked."

Well, that explains the vibe at the school bus stop.

But next, he says something that makes no sense at all. "I know what you're thinking," he says, raising his arms in an "I don't want to hear it" fashion, "but we really haven't had any problems there."

What in the world is he talking about? There are huge problems at the rehab/hotel. We've seen sheriffs and ambulances at Storybook Inn. I know the sheriffs are called there for disturbances. I've seen red and blue lights flashing off the trees that overhang the hotel.

One neighbor challenges the sheriff captain, "I saw a sheriff's car there yesterday."

The captain pauses and seems caught in the act. "Oh. I guess I missed that one."

"Missed that one?" I don't believe him. We've seen the sheriff at the school bus stop many times. I have pictures of sheriff cruisers. I called the sheriff myself when the "don't fuck with us" gang cut all the security cameras off of the school bus stop.

GANGSTERS AND SCHOOL BUSES

WHAT CAN WE DO AGAINST organized crime? I don't know. We're just a bunch of small-town homeowners. But I talk to some neighbors, and we form a plan.

If we can't get rid of the criminal gang, at least we can help parents keep an eye on the kids at the bus stop. So I get out my credit card and buy a webcam so parents can monitor and make sure the kids are safe. But the "don't fuck with us gang" doesn't even let the camera stay up for one night.

The next morning, my phone rings with the manager of Sky Forest Water Company on the phone. "Dave, they cut down the webcam."

Apparently, after everyone in Sky Forest went home for the evening, gangsters broke in, cut the internet line, and climbed on the roof of our water company. They not only cut down the school bus webcam, they ripped out the wires from every security camera on the building.

My blood starts to boil. Kids are coming to the school bus stop to catch morning buses for elementary, middle, and high school. But instead of a webcam to keep an eye on the kids, we have mangled wires dangling from the roof.

This kind of terrorism sends a message to law-abiding neighbors that nothing can be done to stop the bad guys or protect the kids. And I notice this causes a fight-or-flight response in my neighbors. People in Sky Forest are becoming more and more afraid to speak out with each act of unanswered criminality. The sheriff saying "there's no problem here" sends another message. Crooked cops will cover up the crimes the bad guys do.

Now my neighbors know that "criminals" won't be touched by law enforcement. If they're seen standing up to gangsters, they become a target for violent retribution. Our necks can be cut just like the wires on those security cameras. No wonder my neighbors are starting to whisper, just like the business owner, afraid someone will hear them complaining.

Remember that scene in The Godfather where neighbors protested against the mafia? No? Me neither. Anyone publicly protesting against the Mob will end up as a corpse in the river, and police won't even investigate the crime. When gangsters have law enforcement on their side, they can get away with murder.

I call the sheriff's department and fill out a report on the burglary and vandalism at the school bus stop. But I sense that nothing will be done. The fix appears to be in.

One neighbor says he's seen fights and drug deals at the school bus stop. It's a horrifying mess. I sense the sheriffs know more than the sheriff captain said about the school bus stop scam. They keep records on the 911 calls.

But there's another problem. The sheriffs' thin resources are being gobbled up by gang activities. If deputies spend extra time at Storybook Inn, they can't patrol our community. Deputies told me they don't have time for the amount of crime at Storybook Inn when they have an entire community to protect. This is why we have zoning laws: to prevent harmful businesses from hurting residents and endangering children.

I call the Sheriff's Department and ask about cop cars and ambulances I've seen at the school bus stop. They repeat the captain's bizarre assertion, "Sheriffs have never been there." That's a lie. I've seen sheriffs there. This smells like a cover-up.

I've seen three sheriff cruisers at a time with the ominous flashing police lights bouncing off the trees in Sky Forest. This isn't Compton, but it's starting to look that way.

I'm new to politics. I've never seen corruption before. But I can read the writing on the wall. It's written in tattoo ink. I wonder how far authorities will go to cover up the organized crime ring? If one of us is murdered, will authorities even investigate?

I hope the kids at the bus stop don't get hurt. I hope hotel guests aren't killed. But the kids are there every day, and I watch cars pull into Storybook Inn. Tourists park in front of the hotel sign, unaware of the danger.

Letters to the Editor

Drug rehab and citizens clash

To Gail Fry I was an employee there at Above it All treatment centers for 4 months, and I can tell you there is way more nefarious things going on there than what the community is seeing, sexual abuse, mediocre medical care, shoddy business practices and more. It would behoove the community to do more to crack down. RL

DR. MELVILLE

"Document everything..." says Dr. Melville to one of our weekly neighborhood meetings. He says he's the chiropractor that used to be in the school bus stop building. But the "don't fuck with us" gang pushed him out of the location. He says he's been victimized by Storybook thugs who attacked him and his patients.

"I had to call the sheriff when things got rough. I was assaulted," says the frightened doctor.

I guess we're supposed to believe the sheriffs forgot about that call, too.

But Dr. Melville isn't done with the violent gang. He came to help us kick them out of the school bus stop.

"I'll tell you how to defeat this," says Dr. Melville. "Document everything."

I'm told that the mafia doesn't keep written records so their crimes can't be tracked. So maybe Dr. Melville is right. Written records could be a way to bring gangsters down.

Dr. Melville holds up a letter he wrote to our County Supervisor to document his experience of the gang.

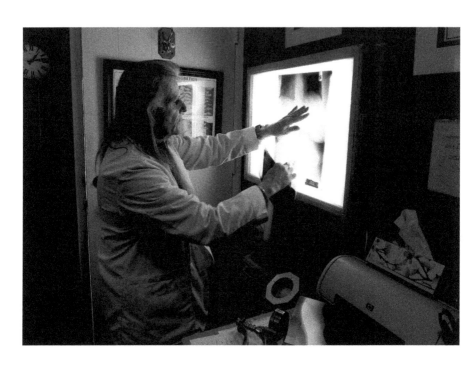

Dear Supervisor Rutherford,

This Drug Rehab business does NOT belong among the public domain.

The inmates/addicts are exhibiting aberrant behavior...

Behavior exhibited in public business building 971 Kuffel Canyon Road,

Sky Forest, California 92385:

1) Masturbation in public restroom.

2) Drawing Phallic symbols on other businesses' windows.

3) Smoking incessantly at my front door.

4) Loitering at my front door 20-30 inmates at a time...Spitting all over pavement.

5) Sheriff and Ambulances showing up to break up fights.

6) Ambulances rescuing inmates who are overdosing...

7) Smoking marijuana at children's school bus stop...

8) Flicking lit cigarettes everywhere...

9) Wild, raucous behavior, acting like on methamphetamines.

10) Flicking ashes onto other businesses' staff (mine) and patients.

11) Rude comments...lewd conduct, aberrant behavior like they are "On something."

12) Calling me, Dr. Melville, "motherfucker" on a regular basis.

11/17/2011, 8:45 a.m.: 5 females in parking lot that faces Kuffel Canyon Road. 1 female is flashing her breasts at the others.

This isn't new. Dr. Melville's story is what I've heard from other neighbors. But Dr. Melville did something that the other neighbors didn't. Dr. Melville put it in writing.

Dr. Melville wrote a letter to crooked politicians who are participating in the school bus scam. He recounts a laundry list of violations. The date on Dr. Melville's letter and eyewitness accounts prove that county officials are aware of crimes at the school bus stop.

MY PRAYER

MY PHONE DINGS WITH A text message. I cringe when I see it's from Verizon. My bill is past-due. I'm sure my cell phone bill is somewhere in that pile on my desk. I guess I better pay it before they shut off my text messages.

But there are other bills in that pile. I have bills for the county's crooked, bear-feeding trash service. I have bills from my little one's mom and her lawyers. I didn't use a lawyer in our child-custody case, but I still have to pay thousands of dollars to *her* lawyers. And then there's a big one: the county's property tax bill.

When Gibbs and I were traveling the country and flying high, I would just write them a check. But Gibbs crashed a helicopter last year and was killed. I don't work as much anymore, so the bills are piling up.

And now a bunch of gangsters have taken over the town. It's almost a Greek tragedy that the same county who allows killers at the school bus stop is sending me a giant bill I can't afford to pay. I can understand why my neighbors are starting to back away from the bad guys. Who can handle the crushing weight of fear and stress under a mountain of bills?

Gibbs used to call me a pea-brain, and he was right. I'm not very smart sometimes. I don't know how to fix this. And I can't see a way to live with what the bad guys are doing at the school bus stop. Our town is not safe. Someone is going to get hurt.

So I get on my hands and knees. On my left is the great old fireplace of Cloud Nine. It was built in 1925 by men who survived tougher times than these. On my right is a pile of past-due bills. The bills will have to wait. It's time to pray. "Lord please help those who enter the Storybook Inn. If they are drunk, help them to get clean and sober. Help the workers to show kindness and love for their guests. Please send us a powerful leader to overthrow this corrupt regime, and help my neighbors get their beautiful town back."

I'm taking Dr. Melville's advice, so I start documenting everything. I take out a spiral-bound note-pad and write down the things I've seen at the school bus stop, including dates and times.

"THERE ARE OTHERS…"

A NEIGHBOR COMES TO ME AT one of our town meetings. "They're having a drug rehab problem in Crest Park, Dave. Maybe you can help them." My neighbor leans in with hungry eyes. She really wants me to go to Crest Park, but I don't know how to help other towns. I don't even know how to fix our town. But I always take advice, so I look up Crest Park on a map and crank up the diesel.

The turbo sings as hundred-foot Douglas fir trees drift by my truck mirrors. I weave right, then left as the winding mountain road leads me to Crest Park. I park in front of a nice-looking home in the forest and knock on the door. I'm greeted by a nice man named Bob, who has a disturbing story to tell. He points to the house next door. "Someone was masturbating there last night." Bob shakes his head in disgust and says that when he yelled for the man to stop masturbating in full view of his children, the drug addict just stared back and kept masturbating. I can't believe what I'm hearing. This disgusting sexual crime took place in full view of Bob's family, in this quiet neighborhood.

Bob describes being "accosted" on his street by drug-addicted thugs. It's the same as our school bus stop. Bad guys are running amuck in between the houses on Bob's street.

"Wow!" I say. "How can this be happening in Sky Forest and Crest Park?"

Bob shakes his head and chuckles. It's the same laugh I get from Hugh when I ask why San Bernardino County authorities aren't enforcing the law. Bob studies me the way you'd watch a dog chasing his tail. He seems to take pity on me and offers, "This isn't going to end, because our county supervisor is in on it." A light bulb goes off in my head. Janice Rutherford is our county supervisor. She's also Louis Murray's boss! Janice Rutherford is the same politician that Dr. Melville wrote a letter to last year.

Bob says he can't get crooked San Bernardino County officials to restore law and order in his neighborhood. It's exactly what I've witnessed in Sky Forest. Time and again the conversation turns to Janice Rutherford.

Bob also says he's aware of the problem in Sky Forest. "People are getting high at the school bus stop," he says. But Bob has been dealing with this for longer than I, and he's giving up the fight against corruption. He says he'll sell his house and move his family some place safe. That's a shame. If good people move away, how can we ever get enough support to fix this?

"OUR HANDS ARE TIED..."

As Bob runs away from the problem in Crest Park, things get worse in Sky Forest.

"Dave, I'm afraid they'll kill us," says a neighbor, pushing nine-millimeter rounds into a pistol magazine.

"Dave, I can't talk to you anymore," says another neighbor.

"Why is that?" I ask.

"They're attacking my business. They'll shut us down if I don't stop helping you. I could lose everything."

I'm having a tough time getting ahold of people. One by one, my neighbors have stopped taking my phone calls. Some have stopped answering texts and emails. One neighbor, who spoke publicly about the bus stop crime ring, suddenly contradicts his own statements about the bad guys. He emails me detailed statements of loyalty to the bad guys, saying the school bus business is "perfectly reasonable." When I write him back and politely question him about his new story, he doesn't respond. In fact, he's never responded to a single phone call, email, or text ever again.

Our weekly meetings are getting smaller and smaller. We used to have thirty or forty people raising their hands to volunteer and shout suggestions at every meeting. But now only a half-dozen neighbors gather in the woods, behind the pizza restaurant. Expressions are not energized and optimistic as they used to be. My neighbors appear depressed and fearful.

"Why did Jerry lie to the newspaper about the rehab's water usage?" asks one neighbor. "He knows they used our entire town's water supply. The mountain spring is dry. We have no water left. Why would he lie about that?"

Hugh takes a deep breath. He shakes his head sadly and says, "They're just afraid. People are afraid of the bad guys." Hugh is an expert on public corruption, and he's seen this before.

Because I'm speaking publicly, and some neighbors are hiding, I'm becoming a target. When people like Bob in Crest Park give up the fight and move away, it makes me even more of a target. If I'm the only one complaining about gangsters, getting rid of me will leave them free to rampage.

I wish people stuck together to fight and make the school bus stop safe again. There's safety in numbers. But I guess that's not what people do. Some people are hiding from the bad guys. I've noticed that some people are also trying to make nice with them and make money off their illegal business.

I'd like to say I'm not worried, but the truth is, the danger in Sky Forest is out of control. When criminals make the rules, any one of us can be killed. I suppose that's why Bob stopped returning my phone calls. He fought this corruption for years in Crest Park, but now he's afraid to answer my phone calls or send me a text message. Anyone can be made a target if they're among the last handful of people standing up and fighting back.

The common thread from one town to another is the name Janice Rutherford. Her name comes up time and again. And when I ask for law enforcement help, county officials say, "I hate this, Dave." "I'd love to help you, Dave, but our hands are tied." I wonder if Supervisor Rutherford has the power to tie their hands...

THE NEWS

BACK AT WORK: John Tamburro pushes the throttle lever on his AS-350 helicopter. The Turbomeca Jet Engine winds up to fifty-thousand RPM. He carefully pulls on his collective lever and we rise off the ground. I flip the mode switch on my Cineflex Laptop and our powerful lens becomes gyro-stabilized. I pull the joystick, and a thousand LA buildings drift across my screen like a moving jig-saw puzzle.

But in one of those buildings, a murder was committed. And that's why we're here. John lowers his collective, and we dive over a cliff toward the Pacific Ocean. "That's where the victim's body was dumped," says a CBS News Producer. "OK. Can we fly from the ocean to those rocks?" he asks. John pulls on his cyclic stick. We circle a lighthouse, dive down until the foamy-white of the ocean waves are just inches below our skids. The picture on my Cineflex screen looks like a scene from Avatar.

As the sun sets, and city lights twinkle in my viewfinder, John lowers his collective and we land after a long day of filming for CBS. I hand the producer a stack of shot tapes, and I tell him about the crimes in my little town in the mountains above us. I figure since he reports on crime for a living, maybe he'll have some words of wisdom.

The producer's brow furrows. He can't seem to wrap his head around the scene I describe at our school bus stop. I don't blame him for being shocked. A bunch of gangsters taking over a school bus stop doesn't sound possible in a country ruled by laws. But the producer makes a phone call and alerts the local station, CBS 2, and a news crew drives up from Los Angeles to do a story about Storybook Inn.

After suffering the "secret business" and street gang, it feels good to have CBS bring the power of their spotlight to Sky Forest. My phone lights up with an Orange County phone number.

"Where is your town located?" asks the CBS2 reporter.

"Santa's Village." I tell him. Anyone who grew up in LA for fifty years, remembers the Christmas Theme Park in Sky Forest.

"Oh, by Lake Arrowhead?" he asks.

"Yes, in the clouds above the lake," I say.

SKY FALLING
A Documentary

"Hmmm. Really?" he sounds perplexed. I don't blame him. How could a crime ring be allowed to set up in a Christmas-themed town at the head-waters for Arrowhead bottled water? Every fridge in America has a bottle of Arrowhead Bottled Water on the shelf. The idea of a criminal gang taking over such a pristine place doesn't seem possible.

"OK. Got it." says the reporter. "We'll be there in an hour. Will any of your neighbors go on camera with you?"

"I don't know."

"Hmmm. It'll be hard to tell the story if people aren't willing to go on record," he says. The reporter's words splash over me like a bucket of ice-water. CBS is bringing cameras to our school bus stop. Thank God. But the reporter is right. If no one is willing to be seen on TV complaining about the gangsters, this might be a waste of time.

"I don't know if my neighbors will go on-record," I say. "That's a problem I'm having right now."

"Please be careful when you drive up the mountain," I warn the reporter. "We're in the clouds today. It'll be like driving through a heavy fog."

I watch out my windows as clouds swirl through the trees around the house. Out of the fog, a brightly-colored news van emerges with its headlights on. I could kiss the cameraman and producer as they pull around the corner and park their brightly-colored news van in front of the school bus stop. A tattooed gang-banger spits out his cigarette and hurries out of sight. It's the first time I've ever seen the gang run away. Normally they charge across the street like raging bulls if anyone takes a picture. But not today. Today, the burly thugs pull their side-ways ball caps down to hide their faces. Suddenly the criminal kingpins of Sky Forest look like trapped animals. Not a single one of them dares walk around outside with a CBS news camera pointing at the place.

Clouds swirl around the Storybook Inn Hotel sign as the CBS cameraman pans through the gangsters' lair. It's hard to see anything in this fog. How appropriate for a secret illegal business that regulators say doesn't exist.

The CBS cameraman focuses his lens on a school bus that stops and lets out elementary students. I hope people see this on television. But there's a downside to this media spotlight. Tonight, when the news van leaves, the gangsters will come out of the shadows. They might seek murderous retribution at any house where the news van parks for an interview. As people become more and more afraid of the "don't fuck with us" gang, my house is one of the few where the reporter can still park his news van.

The cameraman sets up his tripod next to my dining room table.

"This makes us completely unsafe," I say to the CBS reporter as his cameraman zooms in on me.

But my neighbors refuse to appear on camera and tell what they know. People are afraid of retribution from thugs and corrupt officials. They complain for hours about the assaults, graffiti, trespassing, burglaries, but no one is willing to say a word to CBS News. I argue and plead for half an hour to get a single neighbor to go on camera with me.

The reporter sits at my dining room table trying to make sense of the crazy situation that people are afraid to talk about. He calls county spokesman David Wert on the phone and asks, "Is Storybook Inn a hotel or a drug rehab?" Aah. That is a good question...I can't hear the answer, but county spokesman David Wert is quoted in the CBS2 story. "Above it All Treatment is totally in compliance with all county and state regulations."

"What?" I gasp. I've never seen anything more out of compliance. David Wert gave a false statement to CBS. I have a thousand pages of documents that say that Storybook Inn is completely *out* of compliance with state and county regulations. They're using illegal drugs and assaulting people in the presence of school children. They're committing fraud, vandalism, drunk and disorderly conduct — where county school buses park six times a day. The Bates Motel was more in compliance than Storybook Inn.

David Wert's statement contradicts everything I've witnessed at his county's school bus stop. His statement contradicts everything I've heard from his own county regulators. His statement to CBS even contradicts the gang's own nation-wide advertising. Meanwhile, David Wert's county school system orders children to stand in front of the place five days a week.

When I look across the street at the school bus stop, I see tattooed thugs swallowing fistfuls of narcotics, smoking, and shouting profanity. The thugs even cut the security cameras off the building next door so no one can keep an eye on the kids. Our school bus stop is a criminal powder keg. I know someone is going to get hurt or killed. It's only a matter of time.

Storybook Inn owner Kory Avarell tells CBS2 he wants the public to come to him.

What? I already came to him. But his operations manager, Kyle Avarell, said I couldn't talk to him.

But Kory looks charming on TV. He says, so sweetly, "We can have a meeting right here."

That's not true. I already asked for a meeting right there. But I was accosted by thugs and told "this conversation is over."

God bless our friends at CBS news. They caught thugs and corrupt county officials with their pants down. Now we'll see if they're in compliance and ready to meet with the public like David Wert and Kory Avarell claim...

Dr. Melville says, "Document everything." CBS2 helped us by putting the bad guys on record.

ON MY KNEES

As the bad guys lie on TV News, I have little hope for the victims of this crime ring. Government officials say Storybook Inn doesn't have a rehab or hotel license, but David Wert is telling the public the hotel is "completely in compliance."

The sheriff captain says the hotel is run entirely by criminals. My neighbors are terrified to go near the place. Meanwhile, thousands of drug addicts are coming from all over the country, because the pictures look so nice on www.aboveitalltreatment.com. Victims don't know that there's a "School Bus Stop" sign in front of the place.

My knees hit the hardwood floor at Cloud Nine and I press my head against my hands. "Please God, protect the people at Storybook Inn. And send us someone to fight for us and help us get our town back from the bad guys…"

"YOU DON'T KNOW WHAT YOU'RE DOING!"

"You don't know what you're doing!" my friend Hugh chastises me. Hugh has forgotten more about corruption than I will ever know. Right now, he's frustrated because I'm charging into Janice Rutherford's public chambers. I don't have a plan, but I'm heading straight into the lion's den.

The good news is the CBS story helped me gather enough neighbors to stage a little protest during the "Public Comments" section of Janice Rutherford's Supervisor meeting.

"Dave! They *are not* here to help you. They're the ones RUNNING THE SCAM." Hugh drills me. I know Hugh is smarter than me. I know he's trying to warn me out of harm's way, but I can't keep quiet about this the way some people have been doing. I can't live with a criminal gang at the school bus stop. I have to do something. And the only thing I can think of is to create a disturbance in the austere chambers of Janice Rutherford.

God Bless Hugh and my other neighbors who wake up early and take time to jump in their cars and parade to Janice Rutherford's office. I fire up the diesel. The turbo sings as I pull the big truck onto Rim of the World Highway with Hugh and a small procession of neighbors.

The suspension of my old truck groans as we barrel down the forested switchbacks through the granite narrows. The City of San Bernardino sprawls across my windscreen.

As we leave the trees of Sky Forest in my truck's rearview

mirror, the air heats up. Douglas firs and oak trees give way to graffiti and razor wire. As we near Janice Rutherford's office, the fresh mountain air turns to city smells. The desert heat makes me sweat a little as the smell of rotting garbage and marijuana smoke waft through the windows of the truck. So I roll up the windows and turn on the air conditioning. I park the pickup in front of a shiny county building, and we walk inside.

Hugh does his best to coach me along. "OK Dave, fill out that card over there, and you'll have three minutes to talk during "public comments." Try to sequence everyone's comments so you each talk about a different aspect of the school bus stop. You want to try to get a reaction from the crowd."

I empty my pockets and walk through a metal detector. There's a lot of security around Janice Rutherford's office. How ironic that there's no security protecting the kids from the danger she allows at the school bus stop in Sky Forest.

There is a video of us on the San Bernardino County web site from the San Bernardino County Supervisor Meeting, May 7, 2013. If you watch the video, you'll see Janice sitting in the middle of the Supervisors' podium. She looks very powerful and important up there as we explain the deadly menace at our school bus stop.

I brought a list of things to talk about, but Hugh doesn't like my list, so I scratch the list and just talk about the fact that Storybook Inn is an illegal drug rehab.

What Janice Rutherford does next is interesting.

Janice ushers us into a private room. Her staff closes the door. Janice's staff listen to our complaints, and take notes, but refuse to take action to protect the kids at the school bus stop. Instead, they repeat Janice's unthinkable claim that the drug rehab is not a drug rehab. This is done behind closed doors, out of sight... Fortunately, I took pictures. These are for you, Dr. Melville.

Janice's staff have carefully brought us to a place where no one can hear what we are saying. The closed doors are a great way to keep a lid on this scandal. Hugh is right. I don't know what I'm doing. And this crazy closed-door meeting looks like a scene from a mafia movie. I watch as politicians sidestep our questions and break the law to keep the money flowing to Storybook Inn. Right and wrong don't matter in this room. No one can see us in here. No one can hear us begging for law enforcement to protect the kids.

Hugh was right. I have no idea what I'm doing. My cockamamie plan for a public protest was quickly ushered into this room, where no one will hear a single word that's said about the Storybook Inn.

Hugh says that San Bernardino is the most corrupt county in the United States. I don't know. But we're told that the drug rehab is not a drug rehab, and nothing will be done to protect the kids.

Meanwhile, Janice Rutherford's county school system sends children to stand in front of the rehab to wait for a bus. That is where one of the dads, Phillip Romero, notices a tattooed thug staring at his young children. There are a number of thugs milling about smoking cigarettes. They are covered in tattoos of venomous spiders and poisonous snakes. A husky hoodlum sizes up Phillip and says, "You should go someplace else."

Phillip stares blankly at the burly man with tattooed muscles stretching a death-metal T-shirt. Where exactly is he supposed to go to get his children to school, if not the school bus stop? A chain dangles from the gangster's belt. It swings like a lion's tail as the man stalks to and fro in front of the kids. This looks like a scene from a biker bar, but instead of a Harley Davidson, a bright yellow school bus approaches with flashing lights. Phillip's eyes grow wide as the "don't fuck with us" gang crowds around the bus stop.

Back in Janice Rutherford's chambers we stare daggers at Janice and her staff, who insist that Above it All Treatment is not a rehab and there is nothing they can do to protect the children. The closed-door meeting ends with no action taken. Hugh was right. I don't know what I'm doing.

THE STORY AT STORYBOOK INN

I OPEN A NEWSPAPER AND FIND that there's a manhunt on the mountain. A wanted criminal is on the loose. His mug shot is demonic...and strangely familiar. Hmmm. Where have I seen that face?

I copy the bad guy's name into google. It comes back with a Facebook profile. His bio says he's a driver at "Above it All Treatment."

The man's Facebook pictures are chilling. A lit cigarette droops from a sour face. Long, un-kempt hair and a stone-cold expression remind me of the Unabomber. A biker jacket hangs loosely around an iron cross tattoo. Heavily-inked arms dangle from the torn sleeves of a death-metal t-shirt. This guy works at a school bus stop?

I call Janice Rutherford's office again, but instead of a clear explanation of what's going on at the school bus stop, I get wishy-washy, nonsensical answers.

"It's a grey area of the law, you know?" says a member of Janice's staff.

"Grey area of the law?" I gasp. "There are killers at a school bus stop!" I rise from my seat. I can feel my face turning red.

"David, stop yelling at me," the man whimpers.

"What do you want me to do?" I ask. "There are killers at the school bus stop!"

How can hardened criminals have a grey area of the law after the Muffin Lady was shut down by county enforcers for some fine-print technicality? How can the county give easy access to most-wanted men, while putting shopkeepers and restaurant owners out of business? The poor muffin lady didn't hurt anyone, and her kitchen was boarded up. She brought no harm to the children, but her doors were padlocked by the same county officials who prattle in circles about grey areas of the law that allow the tattooed killer on a most-wanted picture to menace the children.

When I ask Janice Rutherford why the illegal rehab can't be padlocked like the Muffin shop and Santa's Village, she says the rehab is not a rehab. She says it's just a hotel, so there is nothing she can do.

"OK," I say. "Fine. Then I'm going to check in to the hotel."

When my neighbors hear my new plan, they panic. "No! Don't do it Dave! They'll kill you!" my neighbors scream. They have a point. So I turn right back to my neighbors. "Who's going with me?"

My neighbors are probably right. I might get killed if I try to check into a hotel run by gangsters. But I can't be murdered if I bring witnesses. My plan goes over like a lead balloon. My neighbors stare at me in stunned silence. NO ONE wants to go near Storybook Inn. Finally, my neighbor Suzee says she'll go.

God bless Suzee. She's a brave lady and a good neighbor.

I strap an action-cam onto my belt. Since Janice says the drug rehab is just a hotel, I'm sure the staff won't mind if I take a video of my visit. Suzee and I leave the safety of my house and approach the craziest hotel I've ever seen.

I know the gang is running an illegal drug rehab, and the "hotel" is full of drug addicts who are trying to get off drugs. So I decide not to walk into the hotel lobby. Out of respect for patient's anonymity, I wait outside and knock on the door.

The pastel, fairytale paint job gives the gangsters' lair a strange vibe. No one is answering the door. I look around the parking lot. Instead of the family sedans and SUVs you'd see at a hotel, we're surrounded by dilapidated cargo vans with rows of seats. According to the sheriffs and Facebook, one of the vans is driven by a wanted man.

Suzee says, "Look Dave. There's someone in there." She points to a window. I lean in to peer inside the Storybook Inn. There is a blur of activity in the room. Thousands of prescription pill bottles are being shoved into Ziploc bags by workers. It looks like a crack cocaine lab in a gangster movie. Workers are over-filling the bags and stretching them to cram more narcotics in. I've never seen that at a hotel before.

Janice's "hotel" looks like a scene from a Quinten Tarantino film. So of course I bang on the window and motion for the thugs to come outside. "Come here. Come talk to me," I say. They freeze. Everybody stares at me in stunned silence. They appear caught in the act. Finally one of them drops a heavy bag of drugs and bolts toward the window. In one fell swoop the curtains are pulled shut. Now I can't see anything.

What happens next is like that crazy moment in Pulp Fiction when the drug dealer runs out half-naked in his bathrobe. I recognize Above it All's CEO Kory Avarell because I saw him on CBS2 news.

Kory was so nice in that TV news story, but today he seems different as he storms out of the hotel demanding to know, "What's this all about?"

He's holding his laptop computer with the lid open, pointing at us like a microphone. I guess he's recording us.

"Hello, Kory. My name is Dave."

"I know who you are," he says angrily. "What do you want?" That's kind of rude for a hotel manager.

"I want to book a hotel room," I say.

"Ummm. We have no vacancy."

Kory dismisses my hotel reservation as if swatting a fly. Now he turns to Suzee and begins yelling at her. "Are you trying to catch me in something?!" He towers over her like an angry troll. I physically move in between them. He leans around me to continue yelling at her, "Are you trying to catch me?!" he yells at Suzee.

"When do you have vacancy?" I persist.

"What?" Kory quits yelling at Suzee for a moment and seems stumped by my follow-up question. "I already said, we have no vacancy..."

"OK, can you look at your calendar and see when you'll have a vacancy?" I ask.

His eyes grow wide.

"I can't." Kory flatly refuses.

I've never seen a hotel manager so unprepared to book a hotel room.

I ask again, "When is the next vacancy?"

Kory seems exasperated by my questions. "I said I don't know!" He searches the tree branches above him for an answer, but the trees aren't helping. "Would you like to come back?" he pleads.

"Yes," I say matter-of-factly.

"We have no vacancy. That's the third time I said it." He looks around for someone to rescue him from hotel guests.

"OK," I say, "Can you show us one of the rooms?"

He stares at me in stunned silence. "No," he finally says.

"Is there a time when you can show us a room?"

He thinks long and hard as if negotiating a tricky contract. "Yes," he finally admits.

"OK. Great," I say, as if we are finishing a normal conversation.

But this is the most bizarre conversation I've ever had at a hotel. Kory seems on edge. His words are short and terse, with occasional outbursts at Suzee. I don't know why he's so hostile to my sweet little neighbor, who's half his size, but it gets worse. Kory hands a cell phone to one of his workers. "I want a movie of this!" he demands.

Kory, I hope you're reading this. You're getting a movie...and a book...

Kory tries to usher us away, but I'm not done with him. I want the public meeting he offered on CBS2.

"When can we have a public meeting?" I ask, the way you would schedule a dentist appointment.

"No. I'm not interested in that," he says.

What the hell? He offered so nicely to have a meeting on TV. But I guess Kory was lying. He flatly refuses to have a meeting.

"What is that?" Kory points to the action cam on my belt. "Are you recording this?"

"Yes." I say, matter-of-factly.

But Kory points to my camera. "I want that! Give that to me!"

"No," I say.

Korey's worker is standing behind me, still recording with his cell phone camera, but Kory appears panicked at the idea that I'll have a recording like the one he's making of our conversation.

Kory points to my camera. "I want that! Can I have that?"

"No."

Kory flies into a fit. It's like watching a toddler melt down over something they want but can't have. I say, "I tell you what…I'll give you a copy of the footage."

Instead of a hotel manager, Kory looks like a drug dealer who's been caught in an FBI sting. But why would he care if I recorded him telling us there's no vacancy? Besides, he already told his worker that he wants a movie of this. They're filming me right now.

But now Kory's done talking. He rudely bolts back into his "hotel" with his half-opened laptop computer and his worker still filming us with his phone camera. The front door of Storybook Inn is slammed shut.

I shrug at Suzee. "That's it, I guess." We start walking back to my house. I feel bad for Suzee. I'm a big guy and Kory's tirade doesn't scare me, but she's a sweet little lady, and I wish she hadn't had to endure his menacing behavior.

On the way home, I catch movement out of the corner of my eye. An engine revs as Kory's pickup truck roars across the street from the Storybook Inn to the school bus stop. Kory steps out. He actually walks into the same office where he offered to meet the public on CBS2. So I shout at him, "When can we have our meeting, Kory?"

Kory comes out of his office. He thinks for a minute and says, "OK." We can have a meeting.

I say, "Great. Thank you." I point to my action cam. "I've been getting threats from your van drivers, so I'll be recording anytime I'm here."

When Hugh hears the story of my failed attempt to check into Storybook Inn, he points at me with an, "Ah hah" look on his face and says, "That video is very important."

"Oh? Why is that?"

"It's evidence that Janice Rutherford is racketeering."

"What's racketeering?"

Hugh smiles at me sympathetically. He's an expert in corruption. And he's right about me. I don't know what I'm doing.

NO SHOW

A LOVELY SIGN APPEARS ON THE bulletin board at the Sky Forest Post Office, inviting us to a meeting at the school bus stop.

"**We hope to see you there!**" — **Above it All Treatment**

That's nice. I hope to see Kory, too. I know a lot of frightened parents at the school bus stop would like to see him.

Finally, we can have a public meeting about the "don't fuck with us" gang and the school bus stop.

> Above It All Treatment Center welcomes S[ky] [Fo]rest residents to attend a community meeting at our facility on Wednesday, May 29th at 7:00pm
>
> This Above It All sponsored meeting will be to facilitate an open dialogue amongst community members and address any potential issues or concerns as we move forward in 2013.
>
> The meeting will be held at our facility located at 971 Kuffel Canyon Road.
> Please use the parking lot staircase to access the meeting room on the second floor.
>
> We hope to see you there!

Wyatt's a good little boy. I never have to worry about where he is or what he is doing. He always stays right by my side. He never wanders off. If I'm busy making phone calls about the school bus stop or doing other grown-up stuff, Wyatt quietly moves to the next room and plays with his toys or draws with his crayons and markers. He'll do this for hours.

Wyatt uses his tiny fingers and hands draw the world the way he sees it. On his construction paper there's a four-year-old image of Cloud Nine, our house. I find myself in stick-figure form standing next to the house. Wyatt patiently swings his markers back and forth to make an image of his mom and Switch the Angry Dog. He points to Switch. "Switch bites!" he says. And Wyatt is right. Switch does bite. So we take steps to keep Switch away from people she doesn't know. When visitors come to the house, we put Switch behind barriers so no one gets bitten by Switch the Angry Dog.

But now there are other animals in Sky Forest who may attack. That's why we're going to Kory's meeting about our school bus stop. I drive my truck around the corner and park in front of Kory's office. There's already a crowd of people gathering. I'm relieved, because it's getting harder to get people to be seen in public speaking out about the gang. Some of my neighbors chose not to come to the meeting out of fear of retribution from Kory and his workers. I can't blame them. There are more people in the illegal business than live in our town, and that could be a problem. If the foul-mouthed, tattooed workers come, they'll outnumber town residents. It's a scary thought but a real possibility.

But this is an important meeting. We need to make the kids safe at the school bus stop again. Talking to Kory is a good first step.

A school bus mom walks by holding a sign that says "Public Meeting Tonight," and her elementary school daughter is right behind her holding a sign that says "Keep Kids Safe."

I asked neighbors to bring signs to advertise for the public meeting, and also to make a statement to crooked politicians and greedy businessmen about what kind of life we want for our children.

A man walks by with a sign that says "Do Rehab Right." Amen to that.

In addition to signs, I also asked my neighbors to bring written questions for the school bus stop rehab. One at a time, they hand me sheets of paper with questions and comments for Kory.

Thanks to Dr. Melville, I document everything. I crank up a portable generator in the back of my truck and I run page after page of questions through a small copy machine and into a file marked "School Bus Meeting."

There are some interesting notes and questions on those pages.

"Women and children are often home alone near your facility. What is being done to protect them from the threats you have brought here?"

"How many people are in treatment?"

"How many clients have criminal and violent histories?"

"Do you administer drugs to clients? Physicians on site?"

"How many former addicts work at Above it All?"

"How many employees have a criminal past?"

"What will you do to protect my children from the negative elements your business automatically brings to our bus stop and neighborhood?"

"Stop dropping your overdosed patients at the ER without paying for them. We want you to pay ambulance and ER bills, not transfer the cost to us."

It's funny that no one asked a single hotel question. Janice Rutherford, I hope you're reading this.

I googled "Racketeering" after my talk with Hugh. Apparently, racketeering means "Dishonest or Fraudulent Business Dealings — usually by Organized Crime."

By 7:00 p.m., Kory and his enormous gang should be here to meet with parents and neighbors. But when I look around, Kory isn't here. In fact, not one person from Above it All Treatment has shown up for the meeting they advertised. I'm flabbergasted. Out of a hundred workers, not a single one of them is here.

No wonder the sheriff said, "They're all criminals." I look at my watch. It's 7:00 p.m., just like it said on the invitation. We're here, but the gangsters are not. I'm furious. I had to arrange childcare for Wyatt so I could be here. Other parents did too. We all made a good-faith effort to participate in Kory's public meeting and help make things safer for school kids. We're at the place and time of Kory's choosing. But Kory and Co. are not here. Instead, the sheriffs move in to kick us off the property.

Remember the sheriff captain who said he's never been called here? Well, here he comes, and he orders us to leave. He orders us off the property, even though Kory offered on CBS News to meet, and we have a written invitation, still hanging on display at the Sky Forest Post Office. My neighbors all look to me in stunned disbelief. One by one their shoulders slump, and they walk away from the school bus stop. They mope sadly toward their cars with homemade signs swinging at their feet.

Hugh frowns on hearing this. "Dave, do you notice how the sheriffs aren't enforcing the law here? But they come running when Kory's illegal business wants something."

Hugh is right. The sheriffs aren't protecting us. They're not defending the kids. In fact, they're taking orders from gangsters and shoving good people aside to keep the illegal business running.

But there's one thing I don't understand. If Kory has the protection of the sheriff, why not just come out and speak to us? Why did he invite the school bus stop parents if he's not going to meet them?

The sheriffs shouldn't have ordered us to leave the meeting place at the exact time of the meeting. The sheriffs are participating in a fraud. If you're shopping for drug rehab, I suggest you carefully consider any advertisement you see for rehab in California.

It felt like we were making progress when Kory agreed to this meeting, but now I feel defeated as I watch my neighbors wander off without any answers or dialogue from the secret, illegal business. But suddenly a thought comes clear to me. "Wait just a minute. I have a written invitation from Above it All Treatment for a meeting at 7:00. So the sheriffs can kiss my ass."

"Hey, everybody!" I shout. "Come back to Kory's office. I want a picture of this."

I invite everyone right back onto the property and over to Kory's office, where he sweetly said on CBS2, "We can have a meeting right here." We snap a picture without him at the exact time of his written invitation—7:00 p.m., May 29th, 2013.

After the meeting, Dr. Melville told me he's been attacked by gangsters. "They stole my wallet and came to my house. They cut the fuel lines on my car and jammed my driver's license in the windshield wiper."

Now I'm worried for Dr. Melville. I guess he's become a thorn in the side of gangsters just like me.

"Why the windshield wiper?" I ask.

"It's a message. My driver's license has my address on it.

They're warning me to keep quiet about the scam because they know where I live."

That pisses me off. The gangsters are hunting down residents one at a time for retribution.

Another neighbor asks, "Did anyone else at the meeting have their car sabotaged? The gang cut my transmission lines."

"Uh oh...What happened?" I ask.

My neighbor explains that his car was sabotaged at Kory's no-show meeting. Like Dr. Melville, he feels that a threatening message was sent, along with attempted murder. But my neighbor doesn't report this to the sheriffs. With the way sheriffs are acting, I can't blame him. The sheriffs have been acting like part of the gang. After our conversation, the man disappears. He never answers another phone call from me. He never responds to another email or text. He's one of many good people who choose not to speak to me again. It's getting too dangerous to be seen standing against the gangsters.

With every person who stops fighting, I become a bigger target for the gang. And I have a small child in the house. My blood runs cold.

I suppose my neighbors who chose to disappear are the smart ones. But sometimes I'm just not that smart. I tend to get myself into trouble, with no plan for escape. No right-thinking person would do what I'm doing in the face of murderous corruption and a horde of most-wanted men.

Hugh said I don't know what I'm doing. My friend Nathan says I'm, "thick as a brick." Gibbs used to say I could screw up a two-car funeral. And they're right about me. This school bus fight is just another one of my ill-planned adventures.

"YOU SHOULD MAKE A DOCUMENTARY..."

"OK. Now fall back behind him," I say to my pilot. My pilot pulls back on his cyclic stick and pushes down on his collective control. I can hear the Allison C20 Turbine change speed as the Bell Jet Ranger III decelerates. My Cineflex monitor frame fills with 80,000 pounds of steel as a big rig rolls a few feet below our helicopter skids.

The big rig rolls into the horizon in front of us, trailing smoke from a pair of giant stacks that poke above the driver's cab like shiny dragon horns. So ends another helicopter shot on *Ice Road Truckers*. I hope we land soon. The temperature is sub-zero and dropping.

As the truck speeds away from us, the pilot sees something he doesn't like. "I think we're in trouble Dave," he says.

"Uh oh."

"Yeah," he says. "Vis is coming down. We might not make it to Dead Horse." Uh oh. If we don't make it to Dead Horse, we'll be stuck in one of the coldest places on Earth with no heat tonight.

My pilot pulls left on his cyclic stick and the main rotors chop away from the Ice Road and out across the frozen tundra. I pan the Cineflex across the icy wilderness. Nothing moves out there. There are no humans in my view-finder—just snow and ice. There's no forest. It's too cold. Trees don't grow here.

I pan the Cineflex to the last light of the setting sun on the horizon. I press the Cineflex zoom lever and the telescopic, gyrostabilized lens reaches across fifty miles of tundra. Fifty miles of snow and ice flow through my Cineflex frame. My phone is playing orchestral music so I can drown out the thunderous cacophony of helicopter vibration and noise. This helps me to remember when to pan and tilt the camera for people who are watching my footage from the comfort of quiet couches in heated homes. But suddenly the music in my headset stops. I look down. My iPhone is dead. I charged it this afternoon, but as the temperature drops, it takes my phone battery down with it.

I hope we make it to Dead Horse. It's getting dark and cold out here. But we're slowing down. As the horizon turns black, the winds are getting stronger, whipping the snow off the tundra and into the air around us. It's like flying through a snow globe.

"I'm sorry Dave. I can't do this anymore," says my pilot.

I look out front and I see the problem. The wind-driven snow is causing a complete whiteout. It's impossible to discern earth from sky. The result is spatial disorientation that's killed many pilots and flight crews in this part of the Arctic Circle.

My pilot steers us back over the Ice Road and sets us down just as the snow becomes an all-consuming whiteout. Every seam and crevice in the Bell 206 starts filling with wind-driven snow. In this weather, the aircraft will be frozen moments after we shut the engine down.

But we're alive. My pilot carefully guided the 206 to a place close enough for our support team to drive out to us. The powerful beams of a Ford F350 flood our cockpit. Our TV crew is here to rescue us, but we can't leave our helicopter here. The next big rig to come in the whiteout blizzard will crash right into us.

"Hey Dave. Can you see behind me with your camera?"

I pull on the joystick and our Cineflex camera points straight back at our skids. I'm careful to never let the camera see our skids while we're filming, but now it gives my pilot a view behind us where there's a side-road we can park on, out of the path of oncoming trucks.

My pilot pulls on his collective and we're flying again, inches above the ground. He watches my Cineflex picture as he flies backward to get us off the Ice Road. He lowers the collective and the 206 comes to rest on its skids.

My pilot rolls the throttle off, and our gas turbine begins a cool-down sequence. As the engine slows down, the cabin freezes almost instantly. My pilot twists his throttle grip all the way off and the engine stops.

"OK, Dave. Get ready."

"OK," I say as I put on every piece of warm clothing I own. It adds almost a foot to my normal diameter.

"Cover up your face," says the pilot.

"Why?" I look around the cabin for my arctic face-covering, but I can't find it in the dark.

"Because your face will freeze when we get out of this helicopter," he says.

"OK." I continue looking for my hat.

I'm lucky I know my pilot. There are few who could fly us through that blizzard in one piece. And I don't know many other pilots who can use my Cineflex to fly the helicopter backward through a whiteout blizzard.

So ends a typical day of flying for movies and television. I'm never quite ready for all the danger and challenges, but somehow I made it through another day.

Normally, I would leave the thunderous, death-defying world of helicopter cameras behind and travel to the peace and quiet of Sky Forest. After cheating death, all I want to do is go home to my quiet life with Wyatt. But organized crime has taken over Sky Forest. So as I step off the plane from Alaska, I turn toward Sky Forest and my mind suddenly fills with dread. I wonder if the gang killed anyone while I was gone.

"You know what you should do? You should make a documentary, Dave," a friend offers.

I'd rather poke my eyes out with a stick. With my current financial strain, I can't afford a new magazine subscription. I don't have as many paychecks coming in the mail as I used to. When I come home from *Ice Road Truckers*, my PO Box is full of past-due notices. My neighbors aren't going to help fund a documentary.

After getting roughed up by gangsters, I can't even get them to take a smartphone picture of the gang at the school bus stop.

But I always try to take advice. So I pull out my least maxed-out credit card and buy a camera to begin documenting the "don't fuck with us" gang.

This is really out of character for me. I try to live within my means. But I'm borrowing money because I have to do something about the school bus stop…And I work in television, so I should be able to make a documentary…

Wyatt smiles big and puts his tiny hands on the new camera.

"Here you go," I say. "You carry the camera to the kitchen."

"What are we doing, Daddy?"

"We're making a Pancake Movie." I call it, "The Pancake Movie," because in our first scene I make pancakes for Wyatt, and I need a name for the film that won't give him nightmares about murders at the school bus stop.

THE PANCAKE MOVIE

"Dave, do you want me to bring my camera to help you?" Gianny Trutmann is a good friend and a Camera Wizard. He has the same Red Camera they used to make Lord of the Rings.

"Sure," I say. "Yes please!"

SKY FALLING
A Documentary

Gianny drives six hours to bring his $100,000 movie camera to Sky Forest. Suddenly our homemade documentary is catapulted into a 4K Movie.

Gianny is a world-class artist. He builds his movie camera and begins capturing images of organized crime in the cloud forest.

We find sheriffs out in the darkened forest.

"What are you doing?" I ask the sheriffs who look suspicious sneaking around in the dark. "Do you want me to turn some lights on?"

"No. We don't want any lights," says a sheriff sergeant. The men around him are holding body bags. They don't know that Gianny is hiding with his movie camera. Like a camera ninja, Gianny silently changes lenses. He flips his camera screen out and he kicks his camera into high gear, switching his Red Cinema Camera from 4K to 5K. His Red Cinema Camera captures twenty-four pictures a second, with enough resolution to fill a movie screen. Corruption never looked so clear.

ANOTHER SCHOOL BUS STOP

A FTER GIANNY GOES HOME WITH his 4K movie camera, I strike out on my own to document the crime ring with the little camera I bought.

Since Kory didn't come to his meeting, I got zero information from him. So I have to do some detective work to find his business locations. It's not easy. The "secret" business has been camouflaged in peaceful neighborhoods like Sky Forest, but I'm able to find some addresses associated with Kory's name. I fire up the diesel and roll into the first address on the list. Right away, I can see something isn't right. It's a business address, but there are no business buildings or parking lot. Instead of an industrial building that you might use for medical care, I find a family house in a quiet neighborhood.

Neighbors say a company called Above it All was thrown out of the house for criminal behavior. The thugs are gone without a trace, except the scars of crime and chaos burned in the memories of next-door neighbors, who recount drunken fights and burglaries in the wee hours of the night—just like Sky Forest.

I walk to another address and can see the tell-tale signs of Storybook Inn—cargo vans full of drug addicts. Two neighbors say they have safety concerns about the house. What kind of company puts a business in a family home?

One neighbor loves it. She says the house is a women's drug rehab. She says the rehab is a good neighbor and rehab staff kindly plows snow from the neighborhood in winter. That's nice.

Another neighbor tells a different story. "The sheriffs are sometimes here around the clock. And the shredders!"

"Shredders?" I ask. What the hell?

"Shredders. There are giant tractor-trailers shredding documents."

I imagine the sight of Kory's tattooed gangsters shredding documents from his secret business. It probably looks like a scene from *X-Files* or *Stranger Things*.

Once again, a quiet neighborhood has been turned into a surrealistic scene.

The distressed neighbor seems relieved that I'm documenting the story. She continues, "I called the sheriff when people were running in and out of the house at all hours. The sheriffs came and said they would do a stake-out on the house." Well, that's not the response I got from sheriffs at the school bus stop in Sky Forest. No stakeout was offered for the kids at our school bus stop.

I'm curious why the response was different here. "What happened when sheriffs did the stakeout?" I ask.

"Nothing. Sheriffs said they would do a stakeout, but they never came."

Ah. Now I get it. That's the same inaction I've seen at our school bus stop in Sky Forest.

She continues, "Sheriffs have been coming out and shining spotlights around the neighborhood, looking for escaped customers."

"Escaped customers?"

"Yeah. People run away all the time. I don't know what's going on."

"I don't know either," I say, "but you're describing the same problems we have in Sky Forest."

My mind is blown. How can such a huge crime ring operate under the watchful eye of the media? Neighbors are dying for someone to come document the story, expose the bad guys, and get this criminal mess cleaned up. But there's no media here, just me and my little camera. Whatever this is, the media seems to be helping to cover it up. After two media appearances, TV and radio producers don't even return my phone calls. My emails go out to them, but nothing comes back.

I look at my odometer, 250,000 miles. I turn the ignition key on the old Ford and wait for the glow plugs to heat up the super-duty cylinders. I hope this truck keeps running. I don't have money to repair it. But for the thousandth time, I turn my key and the International Harvester engine thunders to life. With a quarter million miles, the turbo-diesel sings like a jet engine (with a herd of buffalo running over tin cans.) The suspension groans as the diesel pulls me through a winding mountain road to the next address. I slow down as I approach a forested neighborhood. Out my side window, I spot a worker stepping out of a banged-up car dressed in a medical uniform. She walks into a house with a sign on the wall that says "Avarell" (Kory's last name.)

This is a nice neighborhood, but there's a weird tension in the air. The mood at the Avarell house doesn't feel happy and loving, like you might expect at a counseling business. I don't know how to describe it, except that I feel a negative energy.

I'm curious how neighbors feel about the Avarell house. So I walk to the house next door. Dogs are barking. I feel out of place. Going door-to-door to investigate organized crime is not what I normally do. I feel like I might get bitten by an angry dog or shot by a fearful homeowner. Who knows what I'll find inside this random house? Maybe it's a gangster's house. The secret business could be in any house on the street, and I wouldn't know until it's too late.

Oh, well. This problem isn't going to fix itself. So I knock on the door. A young man answers with a team of dogs barking up a storm. Through the awkward clamor, he figures out I'm asking about the Avarell house. He calls the rest of the family in, and we sit down at their dining room table, where they begin to describe a surreal nightmare. They say there's foul language, fighting, and drunken disturbances around the clock at the school bus stop.

"The school bus stop?" I say.

The man puts his head in his hands and seems depressed. "Yes. This is a school bus stop."

Sky Falling
A Documentary

But it doesn't sound like a school bus stop. Neighbors say they've seen fights, overdoses, drunkenness, and burglaries. They say sheriffs and the fire department sometimes spend hours with their sirens and lights pulsing into the night. The sights, sounds, and smells have turned the peaceful neighborhood into a crime-ridden mess.

"The high school girls at the school bus stop are being taunted by male drug addicts," says one man.

The plot thickens. I grab my cell phone and call the Sheriff's Department to ask about the Avarell house, but the sheriffs say they've never been there, not even once.

"What?" I ask. "Are you sure you have the right address?"

"Yes. That's right. We haven't been called there," says the lady at the Sheriff's Department.

"See this?" A neighbor holds up his smart phone with a picture of a sheriff's cruiser parked at the Avarell house. He walks over to a window and points. "We have red and blue lights flashing, sometimes all night long."

How can sheriffs say they've never been here when neighbors have pictures of them? I'm reminded of our Sky Forest sheriff meeting, where the captain said sheriffs never went to the Storybook Inn. If sheriffs are hiding the truth, what else do we not know? I say, "Do you know what's happening at the school bus stop in Sky Forest?" They shake their heads. They know nothing.

Maybe that's how the bad guys get away with it. The illegal business runs in complete secrecy, without signs, permits, or public notice. No one knows how far the crime ring goes. And when I ask the sheriffs, they lie and say they've never been there. Media is refusing to cover the story, so victims have no warning—they wander into the place, completely unaware that they're entering an illegal business run by criminals.

I ask the frightened neighbors to speak in our documentary, but they all shake their heads. No one is willing to speak publicly. "Dave, they'll ruin our family business," says one frightened neighbor. Once again I see parallels with the mafia. Victims are afraid to talk because they fear that law enforcement won't protect them. I understand. They just saw that the sheriffs are covering up crimes at their school bus stop. Who knows how far the sheriffs would go to keep this place a secret?

One neighbor says he got an additional trash bin to hold the liquor bottles and drug paraphernalia the gang throws in his yard. But he won't speak publicly, because he fears corrupt county officials will attack his family business — exactly what the business owners in Sky Forest told me. A gangster knocked over the trash bin, spilling liquor bottles and needles into the man's yard. The poor man had to get on his hands and knees to pick up beer cans and pot bongs. He then has to pay the county's crooked trash service to haul away garbage from the illegal business. The man says he comes home to find gangsters parked in his driveway. Thugs look right at him and throw trash in his yard. That's the same thug behavior I've seen at our school bus stop in Sky Forest.

I ask if there's anything I can do to help their neighborhood. One lady says, "Yeah. Will you walk over to that house and ask them what they do in there?"

I can tell that she dreads the place and that no one wants to go near the gangster house. So I walk to the Avarell house.

What I find is no ordinary house. A guy in a medical uniform answers the door and panics when I say, "Hi. I'm Dave Arnold. What do you guys do in here?" Suddenly his medical uniform looks like prison garb. He looks around nervously for someone to rescue him. But his coworkers have a deer-caught-in-headlights look and freeze in their tracks. There's an awkward silence as I smile and look inside the strange home. Whatever they do in here, workers seem to know it's illegal. They're just staring at me as if I caught them in the act of something they know they shouldn't be doing.

The outside of the house is nice, like other homes in the neighborhood. But inside is different. It's not furnished like a normal house. Instead of family furniture, it has cubicles and business desks, like you'd find in an office building. A manager comes to the frightened man's rescue and introduces herself as

Danielle. She says, "This house is the intake and medical detox for Above it All Treatment."

"OK. Do you have a business card?" I ask. Danielle struggles to explain why she doesn't have a business card. I think she might be an addict, too, so out of respect for her anonymity, I don't ask her last name.

Come to think of it, no one ever gives their last name at Storybook Inn. There's usually an awkward pause when they introduce themselves where their last name should be. I suspect all who work here are drug addicts and are trying to keep their identities a secret. I wonder if workers here are being taken advantage of by the same corrupt agents that are allowing the school bus stop scam. (You don't have to pay drug addicts very much...drug addicts can't get a job at other places because of the stigma and possible criminal past addiction brings with it.)

Danielle rummages and finds a business card with no name on it so I can make some notes and have something from her company.

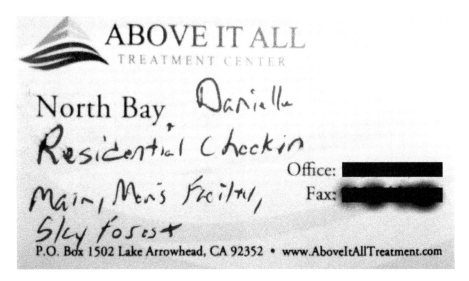

Danielle says, "Everyone comes to me for intake. We process them and do medical detox. Then they go to Storybook Inn in Sky Forest."

They're treating thousands of drug addicts year after year. But how can they fit thousands of patients into a single-family home? Suddenly the nightmare of the neighborhood is explained. The crowds, crime, and chaos are a part of something much bigger than the little house and tiny driveway can hold. No wonder they're parking in the neighbors' driveways and walking through yards. The business in this one small house has more customers and workers than the entire neighborhood — just like the school bus stop scam in Sky Forest. No wonder the neighbors are afraid to go near this house. The criminals outnumber town residents — just like Sky Forest.

As I'm talking to Danielle, I notice there are no signs for her business, just two county street signs painted bright yellow that say "School Bus Stop."

I walk back to the neighbor's house. There I find a house full of hungry eyes and ears. I sense they never go near the crazy gang-house—just like in Sky Forest. Neighbors are afraid to cross the criminal gang. As I sit down at the table, the neighbors lean in with bated breath to hear what happened at the notorious Avarell house. When I say it's a medical intake and detox facility, a neighbor pounds her fist on the table and says, "I knew it! That's against the law!" She complains bitterly, "No one believes me when I tell them how many cars come and go from that house. When I come home, I can't get into my house because the cars from the Avarell house are parked all up and down our street, and their cars are blocking my driveway!"

Another neighbor says that when he complained to authorities, Janice Rutherford came and held a neighborhood meeting about rehab problems at Above it All Treatment. When Janice was done, the rehab was still operating illegally and neighbors became afraid to complain publicly—just like Sky Forest.

"My wife does business with the county. They'll ruin us financially if I say anything," says one neighbor. So that's it. I guess no one outside this neighborhood knows about the problem.

I look down at my notes. Janice Rutherford came here and had dealings with the illegal rehab two years before she told me and my neighbors that the very same company is not a rehab. I'm no lawyer, but I think Hugh is right. This is probably racketeering.

BLACKMAIL

I KEEP PRAYING FOR SOMEONE TO come and rescue us. But so far, no one has come. In fact, the "don't fuck with us" gang is getting bigger, and law enforcement is looking the other way.

A rusting van rattles into the school bus stop. A giant man climbs down from the driver's seat. Cigarette smoke billows from under a sideways ball cap. He stares daggers into me as he unloads a dozen drug addicts at the school bus stop. He doesn't look like the guy from the "Most Wanted" poster, but as he crushes his cigarette under his biker boot, I wonder how many crimes he's committed.

No one I talk to is comfortable with the sight of gangsters at the school bus stop. My neighbors say they want legal action, so we hire a paralegal named Lou Rothman to do some research. This seems like a reasonable step, but it causes some pretty strange phone calls from neighbors. "Don't ever use Lou again!" one lady shouts at me.

I can't figure out why a guy doing legal research is making her so upset. She continues. "Lou is harassing me. He's running background checks on people in the neighborhood." That's weird. We paid Lou for 8 hours of case law research about a drug rehab and school bus stop. Lou was paid in full. What in the world is he doing to the neighbors?

As the organizer of the movement to shut down the illegal business, I now have unexpected responsibilities like dealing with Lou Rothman. So when women in the neighborhood complain that he's harassing them, I assure them we won't hire Lou again.

Besides, I'm not interested in legal action. Lord knows I've had my fill of lawyers. So I'm happy to be done with Lou. But suddenly I start getting messages from Lou demanding money and insisting he's worked for additional hours that no one knows about.

One of Lou's emails sends a chill down my spine. He says a member of my family is a child predator and if I don't pay Lou thousands of dollars, he'll go public with the information. I'm horrified, but not of Lou. I'm horrified for children in the neighborhood. I immediately contact authorities myself, "I've received a letter demanding money and claiming I have a child-predator in my family." A woman at the District Attorney's office listens and says, "OK. We'll walk you through the steps to investigate this."

Next, I call the sheriff's department. When a sheriff's deputy comes out, he asks to see a copy of Lou's letter. The deputy opens an investigation. He comes back and says, "Your family is clear. You don't need to do anything else." Then he adds, "Can I have a copy of Lou's letter? I can't tell you what to write, but I'd like you to keep emailing Lou and let me know what he writes back to you."

If children in my community are endangered, I expect anyone to do what I did, call the authorities and deal with it so the kids can be protected. But Lou offered to keep it a secret for money.

And Lou isn't finished. He posts on Above it All's Facebook page and says he'll provide "free information" to the people at Storybook Inn about me and my neighbors. I think I need a shower. We paid for Lou's time to help make our community safe again — not dig up dirt for the "don't fuck with us" gang.

"Free information?" The only free thing is the time and goodwill I gave to save our town. My neighbors worked countless hours for free. Lou's time wasn't free. He was paid by Sky Forest homeowners to do case law research.

But now Lou claims on Facebook to be providing information to the gangsters at Storybook Inn? I suppose this is the kind of unethical bottom-feeding that happens in a corrupt environment. When the law breaks down, everyone must choose good or evil. And now Lou is posting on Facebook for the bad guys.

But guess what? Now I'm an author and documentary filmmaker. So I don't mind shining a light on Lou right along with everyone else at Storybook Inn.

This is Lou's email.

From: Lou Rothman
Sent: Thursday, December 05, 2013 7:43 AM
To: David Arnold
Subject: Re: I Want The Money You Owe Me Today!

Dave,

I have drafted the lawsuit against you for the $1725 you owe me for my legal services rendered. I will be filing the lawsuit by 11 AM today. The other thing you should know, you are harboring a fugitive that is NOT registered with California Megan's Law for the sexual crimes…committed against that under 14 year old girl…I will have no choice but to notify law enforcement of this fact. I haven't told a soul…however I am about to apprise the community at large that they have a sexual predator that is not being monitored by law enforcement.

You need to text me on my phone before 11 AM, to make arrangements for me to receive $1725 or I will proceed with the aforementioned. You are complicit in the crime as well, I hate to go this route, but I am done "trying" to collect what is rightfully owed to me. I appreciate your immediate cooperation.

Sincerely,
LOU
Text me herein that you will have my money today.

Megan's law protects children by informing parents of the whereabouts of convicted sexual predators. I don't know of any Megan's law violators in my family. But if there are, they must be found and addressed to protect the kids.

I discuss this with my family. Everyone agrees that we'll follow the law.

LOU ROTHMAN

When Lou doesn't get money, he sues me. What happens next is a lesson for anyone thinking of suing me. I walk straight into court. Three of my neighbors kindly make the trip to support me and testify to our hiring arrangement with Lou.

In court, the judge and a mediator ask if Lou can show the legal research he did for the money we paid him. Lou says no. He has nothing to show. The judge stares at him in stunned silence. Did a man who considers himself a legal expert come to court with no evidence?

The judge tries to help Lou by giving an example. Apparently the judge also does legal research for hire and explains how he presents findings to clients with a spreadsheet to show hours worked matched to case-law findings. If you hire this judge, he delivers an hour-by-hour accounting for his time. But Lou didn't deliver case law or accounting. And we already paid him in full. That seems like a bad deal for us.

The judge asks Lou if he has any records he can show the court. Lou says no. He can't show that he did *anything* we paid him for. I'm starting to see why my neighbors aren't happy with Lou's work. The judge seems equally offended by Lou's lack of return for the money we paid.

Now there's the extra money Lou is blackmailing me for. The judge asks Lou to show a written agreement for extra work and fees. Lou says no. He has nothing. Lou has no evidence of work paid and nothing to show he was re-hired. The judge is not amused. Case dismissed, and Lou has to pay court costs.

I'm glad I didn't pay Lou's blackmail demands. Let the chips fall where they may. If you're thinking of suing me, the book, or the documentary, then get ready. I will fight you all the way to a judge and jury. So if you think you can weasel into an easy payday, fasten your seatbelt. I believe people will want to know your role in the school bus stop scam.

I checked on my relative, who Lou claims is a child predator, but can't find records of wrong-doing. But I'm told that something happened in the 1970s involving alcohol and inappropriate contact with a child in my family.

While this is shocking to me, it must be addressed. So I work with the District Attorney and Sheriff. After investigation, they say all is well and nothing further needs to be done. That's good enough for me.

If there can be evil deeds and dark secrets in my family history, then how about Storybook Inn and the thousand intoxicated strangers that are crowding our school bus stop each year? I don't know, but Kory Avarell says drug addicts and alcoholics are "giving up control" in one of his slick rehab marketing videos. That's doesn't sound good for the kids at the bus stop.

"Get moving, Bitch!" a tattooed gangster shouts. He stares daggers at a drug addict who stumbles out of a van, along with a dozen other drug addicts that are assembling at the school bus stop. The gangster flicks a lighter, cigarette smoke forms a sinister cloud above his head. He hands the lighter to a line of addicts who, one at a time, light up cigarettes in a tattooed smoke-chain. The "don't fuck with us" gang has changed our lives in unexpected ways. Sometimes I don't recognize my own town any more.

FACEBOOK

"You should make a website for your town," a friend advises, but I'm not interested. With bills piling up, I don't have time for this. On the other hand, I always take advice. So I follow my friend's suggestion and begin building a website and Facebook page for the school bus stop.

You can visit at www.saveskyforest.com and on facebook.com/SaveSkyForest.

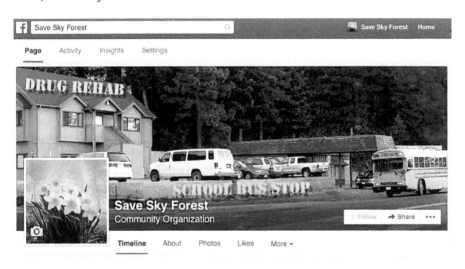

Now we have a place to share information about the secret business. But I found out the "don't fuck with us" gang is on Facebook, too. I see hundreds of foul-mouthed postings on Save Sky Forest in an attempt to scare witnesses into silence. "Who is this? Who's running this page?" a gangster demands. No wonder my neighbors are so afraid. My neighbors stop posting on Facebook almost immediately. The bad guys win again, and the good guys fall silent about the bus stop.

But it's not all bad news. I make a spectacular discovery on Facebook. The gang members on Facebook have been getting arrested. They menace me and my neighbors, and then they brag about going to prison on their Facebook time lines. They don't seem worried about law enforcement at all.

Storybook workers post about who works there, what the schedules are, and when drug addicts escape. Addicts are "escaping?" God help them.

HOME SWEET HOME

"There are guys with big flashlights shining in my windows!" a neighbor says on the phone. She sounds terrified. I don't blame her.

Sky Falling
A Documentary

My neighbor is hiding in her kitchen as gangsters rampage outside her windows. She turns off her lights, terrified they might see her.

We launch a rescue mission from our house but get there just as the rehab van burns rubber with a load of bad guys onboard. The engine roars as the van screeches up her driveway and disappears in the fog. That, my friends, is not the "Spirit of Recovery." That is a van full of killers. The next day, the sun shines on their skid mark all the way up my neighbor's driveway. You can see the picture on: www.facebook.com/SaveSkyForest.

I go back to the gun store and take a number. I wait in line for four hours. It takes an act of congress to buy a gun in California, but I have to do this. I need to protect my family. Criminals outnumber the residents of our little town.

"WE HAVE TO STOP!"

N<small>OT EVERYONE IS BUYING GUNS</small> for protection. Some people choose other ways to deal with the threat.

"Dave, you have to pull the petition!" one of my neighbors screams at me on the phone.

Why is she mad at me? It's not even my petition. But she yells at me to stop a petition drive to shut down the illegal business at the bus stop. Strangely this is the most outspoken neighbor, who's been demanding action from other neighbors. She's the one who gathered signatures for the petition to force county law enforcement to protect us from thugs at the school bus stop. But now she feels she's in danger of retribution and wants me to use my influence in the neighborhood to stop the petition and keep things quiet. She says her business is threatened by bad guys and corrupt officials who will ruin her if she's seen calling attention to the school bus stop scam. She even wants me to contact other neighbors and tell them not to file the petition she asked them to sign.

I'm glad no one can see my face during this phone call. We're talking about a school bus stop. If you're not willing to make a move to protect the kids on your street, then I'm not impressed.

"No," I say. "I'm not telling anyone to stop anything." It's ridiculously hard to get anyone to do anything about the school bus scam. I'm not going to tell anyone to stop what they're doing. Now I have to pull the phone away from my ear. As I set the phone down on the table, I can still hear my neighbor and her husband screaming at the top of their lungs. Not only do they want to stop fighting the "don't fuck with us" gang, they want me to tell others to stop fighting and keep the whole scam a secret.

At times like this, I wish I could quit. It would be so easy to look the other way. But maybe the world needs more stubborn people who can't think of every reason to quit doing the right thing. I prayed for a powerful person to come to our rescue and fight against the corruption that's ruined our town, but nobody's coming. We'll have to do this ourselves.

I didn't expect the people who'd demanded action to scream for me to quit, but here we are. And that is how I became the last man standing up to an organized crime ring.

It's a shame. If we all stick together and fight, there's safety in numbers. If everyone wimps out, a target will be made of anyone left standing against the gang. And the bad guys know I'm alone in this.

Some of my neighbors write nasty notes or yell at me to keep quiet about it because they fear retribution from the bad guys. At first I was hurt by this betrayal, but I toughened up. My dear friend Nathan Crawford said I'm, "thick as a brick." Sorry Nathan, but I think I'm getting thicker.

I decide not to take attacks from my neighbors personally. I think I'm growing up in the process. I decide not to yell back at people who scream at me to be silent about the crime ring. They're just scared. So they're trying to get out of harm's way, even if it leaves the children in danger.

Fear of evil is nothing new. The Bible is full of stories like this.

And there are Hollywood movies about it. The good guys in Lord of the Rings spend half their time trying to get someone—anyone—to fight against the bad guys. In *The Desolation of Smaug*, a little hobbit had to fight a dragon alone because no one else was willing to confront the beast. If you want to know how organized crime keeps going at our school bus stop, that's how. People are afraid to confront it.

NIGHT STALKERS

"**L**ET'S GO, MOTHERFUCKER!" PROFANITY ECHOES through the cloud forest. Screaming at the hotel usually quiets down within the hour—but not tonight. Hooded figures dart through the trees. Heavy boot steps thunder outside my windows. Their flashlight beams scan across my books and pictures as Storybook workers peer into my home and search for their prey. It's like a scene from a horror movie.

Sky Falling
A Documentary

On nights like this, I keep my shotgun loaded with the safety off. All I have to do is slide the pump action and pull the trigger.

"San Bernardino County Sheriff's Department. How can I help you?"

"There are bad guys running through my yard."

"OK. Where did they come from?"

"They came from Storybook Inn."

"OK. We'll send someone over."

But the sheriffs don't come. We call the sheriffs again and they send a deputy to Storybook Inn, where workers say they're hunting an "escaped addict." According to county officials, there's no problem here, but I know something's wrong.

"Someone escaped from the hotel?" God help them.

Eyewitnesses said that Janice Rutherford has been involved with Above it All Treatment for seven years. That's a long time. I don't understand why she's allowing this to continue, or why she claims it's some other kind of business. Supervisor Rutherford knows the county laws better than I do.

This has to stop. But how can I get it shut down if people aren't willing to go on record and tell what they know? How can I get law enforcement to protect the kids at the school bus stop if law enforcement is covering up the crimes?

There certainly is no one to save us. We complained to media, regulators, the sheriffs, and our highest county official. Everyone seems to be in on the scam or willing to turn a blind eye. And Janice Rutherford's county government requires the kids to be there six times a day.

I set an alarm on my phone to alert me when the high school kids will be there tomorrow morning waiting for the bus. I'll go and keep an eye on them. But I can't be there every day.

I drop to my hands and knees in front of my old fireplace. I press my forehead against the concrete. Before I was born, men dug massive granite stones out of the mountain to build this fireplace. They suffered harder times than these. We just have to hold on until someone comes to set things right.

"Dear Lord, please help the patients to stay sober for one more day, and please help the hotel workers to be kind to them. Please send us a rich and powerful person to take on the forces of corruption and evil. And please watch over my neighbors and help them to get their beautiful town back."

You might think it's crazy to pray for the same thing every damn day as things continue getting worse...And it is crazy. But I'm not one to let go of unreasonable wishes. Right now, that's all I have.

LAST MAN STANDING

It's not easy being the last holdout against organized crime. Because I'm speaking in public, I'm an easy target. All the bad guys have to do is get rid of me, and they can keep making money illegally at the school bus stop.

But you know what? Fuck 'em. I'm tired of bad guys scaring the hard-working, law-abiding people of my town. So I paint SAVE SKY FOREST on a piece of cardboard and walk over to the school bus stop. After a few minutes, a man pulls over and swings his door open. "Hey! Get out of here!" he yells.

"No thanks," I say. "Do you know where you are? This is a school bus stop."

The man stares daggers into me. "We know who you are," he says. He mean-mugs me in "This ain't over" fashion, slams his door, and drives away, never taking his eyes off me. So ends a typical day in Sky Forest since the gang moved in.

I walk back to my house and plop down at my desk. I have a foot-high stack of unread mail, unpaid bills, and records from the criminal enterprise at the school bus stop. Sheesh.

Then the phone rings, but I can't see who's on the caller ID. Instead of a phone number, the phone says some weird message I can't figure out. Hmmm. That's weird. I push the slider on my iPhone.

"David here."

"Quit asking questions!" Someone I don't know is on the phone. I don't know who they are or how they got my number. But they continue, "No one is coming to help you."

Then they say the same things Janice Rutherford says: "It's not a rehab. Law enforcement won't help you at Storybook Inn. Don't speak at the county meetings again, or else..." *Click.* The phone goes dead.

It's funny how I can't get a phone call back from Janice Rutherford, but people keep calling me with her talking points. I suppose I should heed the warnings, but I'll let the smart people worry about that. I hope that someday the bad guys will be afraid of the good guys in Sky Forest. I hope that "Or else..." doesn't go like the bad guys expect. I know they don't expect this book. If you have any strong feelings about this story, please share it...I could use a hand getting the story out about this gang.

For my part, I can't roll over and keep quiet the way some of my neighbors did. I know it's not safe to call attention to a secret, illegal business run by the "don't fuck with us" gang. But some days I feel compelled to walk right up to the Storybook Inn and ask them what they're doing in there. The answers are entertaining. It's going to be an interesting documentary...

Jake Capistron is a good friend of mine. He's a helicopter cameraman who sets aside his usual business to drive up to Sky Forest and pick up my documentary camera. We walk over to Storybook Inn.

"There it is," I say. "The strangest hotel in America. If you value your life, I suggest you don't check in there."

"Good morning!" I shout at a man who just stepped out of a serial-killer van.

"Good morning," says the man.

"Come here. Come talk to me," I say and wave him over.

He looks nervously around and crosses to my side of the highway.

"How's business at the hotel?" I ask.

He looks at me like I've mistakenly come to the wrong place. "This is NOT a hotel!" he exclaims, as if he's worried for my safety.

"Really? You just parked your van in front of a hotel sign." I point to his bedraggled van sitting in front of the big Storybook Inn sign.

He stares blankly. I suppose he never noticed the sign in front of his illegal business, but now he's staring at the hotel sign in front of his van. He puzzles over it like a monkey trying to solve a Rubik's cube.

Finally he comes up with an explanation, "Oh…it USED to be a hotel!" he exclaims, as if the answer to the crazy riddle just popped into his head.

"OK," I say. "What is it now?"

He looks at Jake and my camera and suddenly seems uncomfortable.

"Oh...I don't think I should be talking to you."

He backs away nervously.

"Why not?" I ask.

"I, ummm...I'm not authorized," he stammers, and he hurries back to his side of the street.

I'm polite to the friendly driver of the serial killer van, but I know why he's afraid. His bosses are running an illegal business with a bunch of criminals, and probably some crooked politicians. They're making millions of dollars illegally at the Storybook Inn. But I believe it will be shut down if the public finds out, and that will make this man's bosses very angry. Who knows what they might do to him?

He seems like a nice guy. I pray the drug addicts meet him and not the tattooed villains who storm out of the place and shout profanity when they see me across the street. But that's out of my hands. I pray for everyone at this crazy hotel. May the gang members be kind to patients and hotel guests.

For three years I've been documenting the Storybook Inn. I know it's not safe to be seen watching them. Criminals hate cameras.

But even I can't stay out in front of the hotel for too long.

"Let's go, Jake. We have to leave before things get ugly."

"What was that about?" asks Jake.

"Oh. It's just a typical day here since the gang took over." Jake looks as if he doesn't believe me.

I can't blame him. I've been talking to anyone who will listen for three years. But I guess it just seems too crazy to be real. Anyway, God bless my friends like Jake Capistron and Eric Dvorsky, who take time to drop what they're doing and come to my town to run the camera for me.

We make our way back to my house and my stacks of unread mail and unpaid bills. It's a mess. This probably looks like a crazy town to Jake.

As Jake and I put the camera away, there's a knock at my door.

"Dave! I got it!" Suzee comes in and points to a huge stack

of papers she received from California medical regulators. There are hundreds of pages. My eyes glaze over immediately. I can't make sense of this. Fortunately, Hugh comes in and reads every page. He collates it and divides it by subject and date. He says the bottom line of all the documents is that the State Government issued a "cease and desist" order and a 42-page indictment.

But the criminals at Storybook Inn are not ceasing or desisting. It's business as usual. I count a half dozen cargo vans hauling loads of addicts to and from the illegal rehab. State regulators ordered them to shut down the hotel, but the "don't fuck with us" gang hasn't slowed down.

CEASE AND DESIST

My neighbors are even more discouraged when state regulators fail to shut down Storybook Inn. After years of violence, sexual misconduct, drug deals, medical malpractice, and probably some murders, Storybook Inn is open for business. Even with an indictment, nothing changes.

Most neighbors have given up the fight. I don't blame them. Who has time for this? And who wants to stand up to organized crime if law enforcement won't protect them?

Just like the BBQ I couldn't build in my first book, I can't figure out how to fix this. People in Sky Forest say strange things to me nowadays. "Dave, I can't be seen talking to you." "Please don't mention my name." "I'm afraid they'll kill us."

In the early days, my neighbors would email and go to meetings. Not anymore. I walk through police barriers and x-rays alone into Janice Rutherford's San Bernardino Supervisors meeting. It's ironic that politicians have so much security and armed guards keeping them safe inside a billion-dollar chamber built with our taxes. The fat cats in this room don't feel the heat from the thugs at our school bus stop. The Supervisors didn't go thirsty when the gang used up all our drinking water. Politicians sit in plush chairs surrounded by armed guards. No one gets near them without being x-rayed and watched by men with guns.

Tuesday is not my day with Wyatt, but his mom didn't want him today, so I hold Wyatt's hand as we walk through the high-security perimeter and enter the realm of Janice Rutherford. I fill out a card that gives me three minutes to talk. After years of crime and corruption, it feels good to write Storybook Inn on the subject line at a public meeting.

SKY FALLING
A Documentary

I step up to the mic and ask Janice Rutherford why Storybook Inn is open, but Santa's Village is padlocked by county regulators. Santa's Village is the most positive business in our town. Owners spent millions of dollars getting approvals and permits that Storybook Inn doesn't have, but the county still won't let the Christmas park open. Meanwhile, they send school buses full of children to the Storybook Inn, where grownups fear to walk.

When I ask Janice Rutherford and the supervisors about Storybook Inn, they whisper to each other and giggle. Hard to believe? I'm a little bit shocked myself. But after five years of drug dealing, graffiti, and armed robbery, Janice Rutherford laughs.

This sucks. I'm doing what I can, but no one else is here to ask Janice Rutherford about the Storybook Inn. Thank God, Gianny is filming this. Otherwise no one would believe how crooked these politicians are.

For three minutes, I talk about the crime at our school bus stop. "People have died there. Our town is out of water."

In fact, the illegal rehab consumes our entire town's water supply every two days, and Janice Rutherford knows that. I recorded when our water manager gave her the number of gallons used at Storybook Inn. He said, "Janice, we don't have enough water for this." That was two years ago. And now the mountain spring has run dry. Our town is out of water. I have to pay extra on my water bill to cover the cost of piping water in for the "don't fuck with us" gang. We're all paying a price for this scam.

I can feel my face turn red. My hands are shaking, I'm so mad.

I close with a seething proclamation, "Janice Rutherford, it's time for you to come clean and tell the truth about the Storybook Inn."

That's it. My three minutes are up. No one applauds. No one even seems to be paying attention, as I leave the chamber and walk out to my truck.

"How did it go, Dave?" Gianny asks as we leave Janice's chamber and buckle Wyatt in his car seat.

Gianny knows how it went. He was there. But he needs me to tell the documentary audience how it went.

"I think it went well. I said what I wanted." I try to smile as I start the truck and turn into traffic. But something catches my eye. Police lights.

"Uh oh! I think I'm being pulled over."

Gianny takes one look at the cop behind us and quickly hides his camera. I suppose he can tell the difference between law enforcement and dirty cops. We didn't even get a block from Janice Rutherford and my comments about Storybook Inn.

Now Wyatt and his car seat are out of the truck and sitting on the ground. The cop explains that my license is out of date. The truck is towed and impounded. I have to call a taxi to get us home. Shit. The license went out of date last month. The renewal notice is probably sitting in my pile of unread mail. If I see an envelope that I know has a bill, I just add it to the pile of mail I can't afford to open right now.

Then the irony hits me. I came to city hall to complain about an illegal business that has been operating without a license for *three years*, but the second my license expires, I get the long arm of San Bernardino law. No mercy. Not even a warning. I'm shut down, and the truck is locked in an impound yard.

I suppose there are good reasons why people quit speaking out against Janice Rutherford. I've been warned that there's a price to be paid for stepping up to that mic. I take half the money from my bank account and slide it through inch-thick bullet-proof glass. I smile at the clerk. "Can I get my truck back now?" But the clerk doesn't smile back. She stares at me like a piece of meat and gives me a list of more things I have to pay and do before she'll give me a letter I can take to the impound yard.

Even after paying San Bernardino, I still can't get my truck back. And there is still the towing/impound bill, which will be over a thousand dollars.

STREET GANG

It takes most of the money in my checking account to get my truck back, but I still have bills to pay. I swing a tiny brass door open and peer inside my Sky Forest P.O. Box. Maybe there's a check in here. But as I thumb through a stack of junk mail, bills, and past-due notices, I find no money—just more bills. Oh well, I'll just pay the mortgage this month. All the other bills will have to wait.

My phone vibrates. The bank is alerting me that my account is dangerously low. Thus begins the monthly game of "Which bills can I wait to pay, and which bills automatically pay from my bank account?" If I do the math wrong, the balance goes negative. I get another warning from the bank and an insufficient funds fee for every bill that deducts automatically. The late fees and bank penalties cause a reverse avalanche, where the less money I have, the more money I have to pay to keep things running.

I hope some rich, powerful person will come to save Sky Forest. I don't have the resources to fight city hall.

Three days later, I turn the key on my old Ford at Cloud Nine, but I get nothing but the sound of an eighteen-year-old electric starter turning a dead engine. The truck won't run. Wyatt and I watch a massive cloud of smoke billow around the passenger cab. Wyatt is starting to shiver in his child safety seat. If the truck doesn't start soon, I'll have to take Wyatt back into the house to warm him up. My little one can't sit in the cold for too long. But the old diesel doesn't like to start when it's cold. I turn the key again. *RRRRRR-Poof*—more smoke. Wyatt and I share a look. "Uh oh."

I hope the truck starts. After paying the traffic ticket, the towing company, the impound yard, and the mortgage, I don't have any money left for truck repairs. "I'm cold, Daddy," Wyatt whimpers. I look down at him. His little nose is turning red and his tiny frame is trembling.

"OK. Buddy. Let's try it one more time."

I turn the key, but now the key is stuck. The truck is almost twenty years old. I bought it on Craig's List. And I think the ignition key is a copy of a copy of a copy. Through the years, the copy key edges have worn down and now it's stuck in the ignition. The worn-out key doesn't have enough metal left on the edge to turn the starter switch.

I take a deep breath. A frosty cloud comes out of my mouth and sticks to the windshield. Maybe if I wiggle the key I can get it to turn the switch. A little bit in, a little bit up, and if I hold my tongue just right—*RRRRRR-Bang!* The engine turns and a single cylinder in the old diesel lights. The truck shudders as the single cylinder drags seven others along with it. For a moment Wyatt and I bounce in our seats as the sleepy engine wakes from a winter's nap. Instead of 8 cylinders running up smoothly, 7 dead cylinders coast in silence as the firing order comes back to the one good one. Bang, the live cylinder lights again. Wyatt and I bounce with each blast of the single cylinder. But I'm happy because the rattling, bouncing engine is alive.

Boom-Boom! another cylinder lights. We now have two out of eight. *Boom-Boom-Boom*, another cylinder lights. A cloud of blue smoke completely engulfs the old truck as one by one the heavy diesel cylinders come alive.

I sit motionless as the engine smooths out and the haphazard bouncing is replaced by a thunderous rhythm. "OK, buddy." I pat Wyatt on the back and reassure him. "The heat is coming on soon."

As my problems in Sky Forest get worse, it feels good to have little miracles like a running engine and heat that will come as soon as the engine warms above freezing.

I slide the super-duty shifter, and the old truck surges forward. Wyatt and I bounce through the forest and the cacophonous diesel hurls us over dirt and rocks onto Rim of the World Highway.

With the heat on and Wyatt warming up, we start the winding mountain road through Sky Forest. It's like descending in a helicopter. Wyatt swings left and right in his car seat as I turn our old truck through the switchbacks. Oak and fir trees drip water on our windshield as the clouds stick to their leaves and condense into "tree rain."

I carefully step on the brakes as we exit the cloud forest and approach the first traffic light in the city. Two of my tires are bald, but 10-ply truck tires are expensive, and I don't have money for new tires. So I have to gingerly bring the big, heavy truck to a stop. The desert city streets are warm and dry as a bone, so it's easy to slow the truck down without sliding. But there's a new problem — the old truck roof.

As we slow for the red light, a half-gallon of freezing tree rain comes out of the hole in the ceiling and drenches me on the left side. I let out an involuntary scream, about four octaves above my normal voice, as the icy water runs down my arm into my lap. This causes a hysterical laugh riot from my toddler son, who has been warmed by the truck heater and now sits comfortably in his car seat as freezing rain soaks me to the core.

For a moment I'm furious, as the shock of cold water takes my breath away, and my own son laughs at my misfortune. For a moment I just stare at him and wait for the cold shock to subside. But the contagious toddler laugh riot next to me eases my pain. I can feel myself smiling. I look down at Wyatt, who's laughing so hard he can hardly breathe. It is funny. My crazy choices have left us in an old truck with bald tires, a hard-starting engine, and a roof that sometimes lets water into the ceiling, where it forms a gallon that waits above my head until I press the brakes at this stop light. Now Wyatt and I are both laughing hysterically.

But as I turn our truck onto the freeway, a new problem approaches. And there's nothing funny about it.

A pair of angry headlights zoom into my tow mirrors. The car charges like a bat out of hell, sling-shotting around us and swinging the way you'd pull your fist back in a bar fight. The driver slams his wheel to one side and pile drives into our front quarter. There's an explosion of shiny plastic and metal sparks as the man's car impacts us in the worst possible place, my steering tire.

"Did you feel that?" I ask Wyatt. Parts from the man's car fly over our heads as he leans on his steering wheel, trying to push us into a barrel roll. But the old Ford doesn't give an inch. Thank God. Our clunky old truck weighs three times as much as the sedan that's trying to run us off the road. Thank God for that. After laboring for half an hour to get my old truck started, the massive old truck now protects us from the killer.

"What is it?" Wyatt asks. The truck is so high Wyatt can't see Mad Max outside.

"That car just hit us, buddy. Hold on!"

The man's car is smashed, and he's coming back to hit us again, in a Moby Dick-style torpedo run. Thank God Wyatt and I are wrapped in eight thousand pounds of super-duty steel as we careen down I-210 with a homicidal maniac trying to kill us.

I tap 9-1-1 into my phone as fast as my fingers can manage.

"911, what is your emergency?" says the operator.

"A guy just hit us with his car!" I yell. "Whoa! Hang-on! He's going to hit us again!" The killer swings his car like a wrecking ball. I brace for impact.

"Watch out, Dad!" Wyatt yells. "Why is he doing that?"

"I don't know, buddy. Sometimes people get angry when they drive."

Fortunately, hitting our old diesel truck with a passenger car is like hitting a brick wall with a tomato. The man's sedan is smashed, and Wyatt hasn't moved in his child safety seat.

I prepare to grab my Smith and Wesson as the man circles us like a hungry shark. Thank God Ford built a bomb-proof truck in 1999. The Garrett turbo sings like a jumbo jet as the man smashes into my front tire. And the massive old truck doesn't give.

My front tire is attached to a solid front axle that weighs half as much as the sedan. So all I feel is a light tap on the steering wheel. Ramming the noisy old truck with a passenger car is like punching a statue. The car is crumpled, but our truck hasn't even slowed down.

Suddenly, our would-be killer rockets across four lanes and careens into an off-ramp in a mad-dash getaway. I give the 911 operator his description and license plate number.

"Is the truck safe to drive?" she asks.

"I think so. It feels ok."

"Good. Let us deal with that guy. You can come in tomorrow and fill out a report. We'll keep a lookout for that car."

So we drive to Cloud Eight. Thank God the industrial-strength truck feels none the worse for wear as I turn up the rutted dirt road. I take my foot off the gas and press the brake as we rattle up to our trailer. I pull the diesel down to idle and wait two minutes for the Garrett turbo to cool down like an airplane engine.

I climb down from the big diesel to survey the damage from our would-be assassin, but I can't find anything. Even though the killer's car was smashed to hell, our truck only has a few scratches on the brush guard. If we were in a passenger car, we probably would have flipped upside-down when the man hit our front tire with his car. I run my hand around my balding tire, but it feels ok. Our heavy truck and 10-ply industrial tires saved our lives.

I fill out a crime report for the Highway Patrol, but I get a feeling that nothing will be done. This murderous attack was just a typical night in the big city.

In our RV, Wyatt snuggles himself into the blankets and drifts to sleep. All is quiet after our near-death experience. But dark thoughts run through my mind. Since the "don't fuck with us" gang took over Sky Forest, we've had burglaries, gun and knife assaults, and death threats, and now attempted vehicular homicide. I've heard stories from terrified neighbors who won't even allow their names to be mentioned in this book for fear of reprisals.

I've faced death many times in my career. I try not to think about it. I just go on, doing my job in spite of the danger. But this time Wyatt was there when death came calling. Our truck saved us, but what will happen tomorrow? I don't know.

If I stop fighting, we might not be murdered, but with criminals in charge of our town, will we ever be safe? I don't know. The only thing I know to do is to restore law and order. That means someone has to stand up to the "don't fuck with us" gang. Someone will have to fuck with them until they're gone.

I wish someone else would take up this fight so I can go back to my quiet life. Maybe a movie star or FBI agent will find out about our school bus stop and come to my rescue. My poorly-planned fight against city hall is not going well.

MUDDY WATERS (I'M NOT LEAVING)

WHEN WYATT AND I RETURN to Cloud Nine, I turn my camera on to capture his four-year-old thoughts. I love hearing about the world through his toddler eyes. Wyatt's tiny mind echoes the themes he sees in cartoon movies. Sometimes he speaks in comical ways. And sometimes, like those Disney classic movie scripts, he hits the nail on the head. Sometimes my four-year-old son explains perfectly the world of grown-up corruption.

"Dad? Did you know people are aminals (sic)," he says and holds his tiny hands in the air.

"OK buddy," I say. "How are people animals?"

He shows me one hand and then the other. "People are predators and prey."

How in the world does he know that? Maybe he got it from a nature documentary, or maybe he's overheard too many of my phone calls with City Hall. I don't know. That's why I record these things—because in this moment, Wyatt is right. People are aminals.

I turn the camera off and put Wyatt in his pajamas, and I triple-check to make sure our doors are locked. My door locks offer a thin veil of protection from the "aminals" at Storybook Inn.

But through my wooden door I can hear angry shouts and guttural sounds. I don't know what's happening at Storybook Inn tonight. Maybe the hotel workers are hunting one of their guests.

I've run into these "hotel workers" in the daylight. Their tattooed veins bulge and pirate beards bristle as they shout, "If you don't like it, then move!"

How can gangsters tell a law-abiding citizen to get out of town? Strangely, the authorities are helping them. A thinking person would run from this perfect storm of corruption and murder. But I can't run. I can't leave my home to the "aminals."

In *Book One*, I told the story of a man who refused to leave his home after it was flooded with sewage from Hurricane Katrina. A military helicopter hovered above him as I captured the scene with our Cineflex camera. All he had to do was reach out to accept the helping hand offered by a military flight crew. They were waiting to lift him to safety with their million-dollar helicopter. But he said, "No." He refused to leave his home, even though it was completely submerged in muddy water.

A force of nature took everything from that man in the flooded streets of New Orleans, but he refused to accept it. He stood in waist-deep filth and fought with a hurricane. He refused help when it was offered. He just would not accept the reality that was all around him. I thought he was nuts. But now I have to admit, I'm just like him. At least that guy in New Orleans had US government troops in million-dollar aircraft trying to pull him to safety. There's no help coming to me in Sky Forest—quite the opposite. People are turning their backs on me. The flood of crime and corruption is getting worse.

STRANGE DISAPPEARANCE

Word on the street is that over forty people are missing from Storybook Inn. How can so many people disappear without consequence? I've never dreamed of a situation like this.

But the State of California says Storybook Inn is an illegal business that the sheriff captain says is run by criminals. Owners of Storybook Inn are advertising nationwide for the most vulnerable patients with generous health insurance policies. Some addicts commit crimes and alienate loved ones to feed their addiction, so those loved ones might not travel thousands of miles to California to help when something goes wrong in rehab. I know things are going wrong in rehab because of the screaming that echoes through the forest every time a "customer escapes."

If a drug addict falls in the forest, but their loved ones aren't there…did it happen? If a patient is killed, will anyone even know?

Janice Rutherford is the top county official. She says the rehab doesn't exist.

The sheriff department says they've never been here.

How can anyone know if a person has checked in or checked out?

The Bates Motel looks like a safe space compared to the Storybook Inn.

I think some families will throw up their hands instead of traveling thousands of miles to search for their drug-addicted loved ones. I believe sick, vulnerable patients can be made to disappear if it keeps the money flowing to Storybook Inn.

"Dave! I just saw workers fighting with a patient at the hotel! I'm afraid they're going to kill him!" one of my neighbors says through tears on the phone. Unfortunately, this chaos has become a common occurrence since criminals took over our town. If they don't even have an occupancy permit, how can anyone keep track of the customers?

Some patients try to escape on foot, dragging heavy suitcases. I see them through the windshield of my truck as I wind my way up the mountain. More than once I've rounded a bend to find a hapless drug addict dragging a suitcase in the middle of traffic. There's no sidewalk on Rim of the World Highway.

One patient I talked to fled from Storybook thugs with nothing but the clothes on his back. I don't know how it ended for him. If you've ever seen the gangster movie *Goodfellas*, you have some idea of what it's like to deal with angry Storybook workers. The "hotel workers" look at me like I'm a piece of meat. I wouldn't want to be alone with them in the forest.

All I can do is call the cops when I see gangsters running across my lawn with murder in their eyes. That's how one of my neighbors finds an escaped customer hiding under her stairs. He's hiding from the dreaded "don't fuck with us" gang. His eyes dart left and right, scanning both sides of the street for the serial killer vans.

My neighbor has raised children and grandchildren. She sizes up the scary situation and asks the man what he's doing. The frightened man describes himself as a hostage of "the hotel."

The man doesn't know what day it is. He doesn't know what month it is. But he knows how much they're charging his insurance company to keep him against his will at Storybook Inn. "It's horrible," says the frightened man.

"How about when you wanted to leave?" asks my neighbor.

"Yeah. I wanted to leave, and they didn't want to like, let me go. They were trying to like, hold me as a hostage. It's weird," says the nervous man. He talks like a prisoner of war.

"...You felt intimidated?" asks my neighbor.

"Yeah...they were blocking the door. They took my phone away...I've been here over a hundred days..." says the man as he nervously watches the forest for approaching gang members.

"Nobody chased you?" asks my neighbor.

"Yeah. They were trying to...That's why I hid right here," he points to my neighbor's stairs. My neighbor has a dangerous choice to make. Does she go back inside her house and lock the door, or does she risk being seen helping a victim of the "don't fuck with us" gang? She takes pity on the frightened man and lets him use her phone to call his family to come rescue him from Storybook Inn.

The "don't fuck with us" gang collects over a thousand dollars a day to keep this man in their illegal rehab. He doesn't take his eyes off the road for more than a few seconds in case a van careens around the corner in hot pursuit. One of those vans left a thirty-foot skid mark on my neighbor's driveway when another patient "escaped."

No one involved in this scenario is willing to be identified or go on record and speak about what happened. I don't blame them. Blowing the whistle on organized crime can get you killed.

As darkness falls in Sky Forest, I get on my hands and knees and ask God to protect my neighbors. I hope the drug addicts make it back to their families sober and alive. And I pray for the gang members. May they find mercy in their hearts. May they be kind to the customers for which there are no county records.

"THEY'LL KILL YOU..."

B ACK AT THE SAN BERNARDINO County supervisors', I empty my pockets for armed security and x-rays that lead to the microphone in front of Janice Rutherford.

"Aren't you worried they'll kill you?" asks a man who hears my talk about Storybook Inn. A look of grave concern comes over the man as he leans in to avoid being overheard. "You know they arrest people who speak at that mic?"

"I know. They towed my truck last time."

He looks at me as if he's seen a ghost. His eyes dart around the room to make sure no one else can hear us. "They burn people's houses down."

The man is right. I can't stand in front of the hotel or the school bus stop with my documentary camera for more than a few minutes because thugs will follow me to my house, and Wyatt is there. If they burn down my house, I'll lose more than my house.

That's why I've been keeping the documentary a secret. My neighbors don't know that I spent all my money to film a documentary and write this book.

If you like the book, please share it with your friends. I could use a hand getting the word out.

HELP FROM NEW YORK CITY

"How's it going, Dave?" My friend Charles Ricciardi is a New York documentary producer. His latest documentary is in theaters around the country. It's called "Drew, the Man Behind the Poster," featuring Steven Spielberg, George Lucas, and an artist who painted the most beloved movie posters of all time.

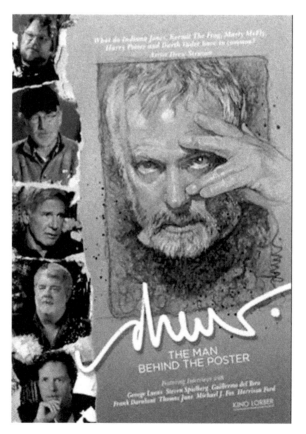

Charles asked, so I tell him, "My town has been taken over by organized crime." Charles looks as if I'd just told him a joke that isn't funny, but I shrug. I'm not joking. Charles is now staring at me, mouth open. For a few seconds he's speechless. He takes a deep breath and asks, "What can I do to help?"

I'm thinking about it. Charles makes real documentaries, not the homemade sort I'm shooting. And Charles is in New York City.

That gives me an idea. "Will you call Janice Rutherford and ask her for an interview?"

"Why would I do that?" he asks.

"You're a famous filmmaker. She'll fear you. And it will put her on notice that if I get murdered, she's being watched by someone outside Sky Forest."

Charles doesn't get it, but he wants to help, so he agrees to contact Janice Rutherford for the interview. And just like that, my homemade documentary has a heavy-weight behind it. And Janice Rutherford's office has to deal with someone other than me.

Charles immediately elevates our game and organizes the filmmaking process. "Do you have releases?" he asks.

"What?"

"Releases. Everyone who appears in your documentary has to sign a release."

"No. I don't have releases."

Charles explains that real documentary film producers like him have participants sign a contract that legally commits them to appear in the movie. But my home-brew project has no such thing. Charles listens patiently and cautions, "You know? It's great while everyone is helping and getting along. But you should prepare for things to get ugly. The people who help you today could attack you tomorrow. If you don't have releases, you can't use the footage. From now on, I don't want you to film anyone unless they sign a release."

So I reach out to a producer friend and she kindly gives me a contract template. Overnight, my friends turn our home-made film into a real documentary with contracts and representation in New York City.

BOOTS ON THE GROUND IN SKY FOREST

"**D**AVE, I WANT TO COME up and help you." Gavin Brennan offers something I sorely need in Sky Forest—boots on the ground. Gavin is a good friend and he's the Director of Photography for *Ice Road Truckers*. He's tough as nails. He drives to Sky Forest and grabs my little camera. Gavin and his cameraman David Mickler leave my house and begin filming the Storybook Inn.

But criminals don't like cameras. They storm across the highway from the school bus stop. Gavin doesn't know these guys, but I do. One of them is Kory Avarell.

Sky Falling
A Documentary

Gavin tenses up as the gang approaches. "I think we're going to have some confrontations with some people," he warns Mickler. "Stay rolling. No matter what happens, stay rolling!"

Gavin is a savvy producer, and he quickly pivots in front of the school bus gang. "Hi Guys!" Gavin waves to the assembling gangsters as they surround Gavin and Mickler. These are the same thugs that threaten me as I walk down the street in Sky Forest — the notorious "don't fuck with us" gang. Vans roll in and gangsters yell through open windows. People have complained they think the vans are used to run over people and haul dead bodies. I'm afraid some hotel guests have disappeared after riding in these vans.

If Gavin and Mickler aren't scared by now, they should be. One of the van drivers is on the most-wanted list. A dad from the school bus stop said thugs tried to run his little girl over with one of these vans. The frightened father had to lunge and pull his daughter to safety as the van driver stepped on the gas. Now the driver steps on the brakes, and the van stops in front of Gavin and Mickler.

"Hey guys! What's going on?" Gavin politely acknowledges the growing horde.

"We're watching," growls one of the thugs.

Gavin is a smart producer who's filmed in war zones and the Arctic Circle. He deftly grabs his phone and fakes a phone call, as if talking to someone who is on their way to meet him, "Yeah. It's all good. Great! See you in a little bit." Gavin cleverly plants a seed in the gangsters' minds that help is on the way.

Gavin speaks matter-of-factly to the criminal horde, "Does the school bus always come on time guys?" — it's a bold move and a way to relieve the tension that's building in Mickler's viewfinder... the criminal gang surrounding Gavin and Mickler is the reason grown-ups are afraid to go near the school bus stop.

I hate this criminal gang. But I have to keep my head down around these guys, because my four-year-old son lives with me. And Wyatt is even more vulnerable than the kids on the yellow bus that's approaching Kory's building.

But Mickler is having a tough time getting a shot of the school bus.

Kory and his thugs dog-pile in front of the camera to block Mickler's view of the bus offloading children in front of the illegal business.

"Aww. C'mon guys," Mickler complains as they push and shove to hide the school bus from his camera. But the dog pile is the truth. Mickler not only captures the school bus, he also captures the bad guys, trying to cover up this unthinkable mess. Thank God for these images.

Mickler picks up the camera and side-steps the thugs to get a shot of the school bus and the kids. Kory Avarell shoves in front of him. God bless Mickler. He manages to get the shot of a school bus offloading children in the midst of a bunch of hoodlums.

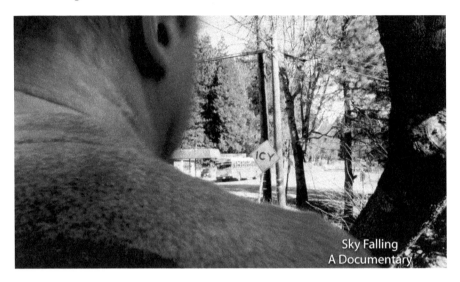

"We can have a community meeting right here."

—Kory Avarell

That's right. The same Kory Avarell who said he "wants the community to come…with their concerns…" now shoves the camera lens away to hide what state and county regulators call an "illegal business" from my documentary camera.

Now I ask you, does this sound like healthcare? Does it seem like a hotel? Does this seem like a safe situation for kids on the bus?

The bad guys think they've hidden the truth from our camera, but they underestimate Gavin, who encircles them with his iPhone camera and asks them why they're blocking our camera's view of a school bus full of children.

Although I've prayed for some rich and powerful person to come and rescue us, no one came, just my friends, who picked up my little camera and walked straight to the school bus stop.

When the school bus leaves, Kory and his gang walk back to the bus stop. Kory plops down in the same chair we saw on CBS2. He was so different on TV. "If anyone has a concern, we'll just take them one at a time." That sounds wonderful!

Gavin catches his breath. "That was scary. Those are big guys!" I don't blame Gavin for feeling scared. My neighbors are terrified of those guys.

Gavin and Mickler got the real story at Storybook Inn, and they got a taste of the threats and intimidation we've lived with since the "don't fuck with us" gang moved in. Gavin and Mickler can go back to their homes at the end of the day, but Wyatt and I have to live with the same thugs that tried to stop them from filming the school bus.

CHARLES RICCIARDI

"**A**M I SAFE? I HAVE a family here." Charles is on the phone and he wants to know if he's endangered by asking questions of Janice Rutherford. I guess he's heard enough about the corruption in San Bernardino to know this will get ugly.

"I don't know," I say. "I hope that by shining a light on Janice, we're safe, like whistleblowers." It's not very reassuring, but it's the best I can do.

My hat's off to Charles for what he does next. He politely, but persistently, demands answers from Janice Rutherford. It turns out that Charles is a tireless fact-finder, asking Janice's office for answers as he pours over records.

I'm surprised that Janice's staff agrees to an interview. They're touting the same talking points Janice gave me and my neighbors — that there is no drug rehab at the school bus stop.

"She better have one hell of an explanation," I say.

Charles scrolls through pages of county code and says, "We need to get a lawyer to talk about the law…"

"OK," I say. It's a good idea. Charles is a real documentary producer, not a shoot-from-the-hip loose cannon like me. And I'm no lawyer. But I can read the three words that tell us exactly what this is in County Code. "Use not allowed."

I'm glad because someone besides me is asking Janice Rutherford for answers. Janice's staff confidently says they only need three weeks' notice for her to sit down on camera and talk about the "David Arnold Issue" and Above it All Treatment. I'm looking forward to this. I'll be amazed if she says the rehab is not a rehab in front of a renowned New York documentary producer.

	TUP	Temporary Use Permit required (Chapter 85.15)
	—	Use not allowed

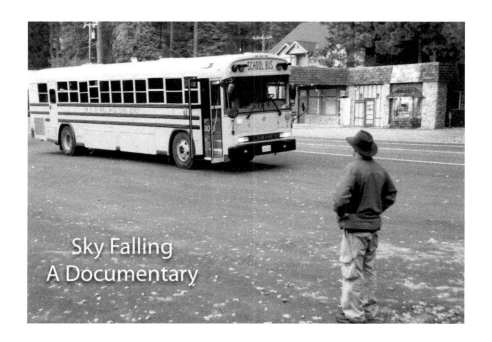

MURDER, INC.

GIANNY LOOKS LIKE A VIRTUOSO musician as he carefully removes a fifty-thousand-dollar Angenieux Movie Lens from a shiny case. He locks the lens onto his Red Digital Camera with the precision of an army sniper, pressing the record button to capture Ultra-High Definition images of Wyatt playing in the daffodils.

I wish I could afford to pay Gianny for his world-class cinematography. But I can't. I have no money for this. When I apologize, Gianny interrupts me. "Dave, I bought this camera to make the world a better place. Don't worry about paying me."

But something catches Gianny's eye. He freezes and whispers nervously, "Dave! Is that the place?" He motions to thugs who pile out of a cargo van and give us the evil eye.

"Yes," I say. "That's it."

"I didn't know it was that close!" Gianny's jaw drops as the violent gang mean-mugs us with their sideways hats and prison tattoos.

"We gotta go, Gianny. Those guys don't like cameras," I say.

We quickly grab Wyatt and the movie equipment as one of the thugs approaches.

"Good Morning," I say, and I tip my hat to the tattooed thug. But he doesn't answer; he just stares angrily at me. His eyes dart to Gianny's movie camera. Criminals don't like cameras, especially when they're caught in the act of making money illegally. We have to grab Wyatt and the camera and go before this gets ugly.

The look on Gianny's face says it all. I don't think Gianny believed the gang was as bad as I said. No one believes me when I tell them about the crime and corruption in my sleepy little town.

I duck down and quickly move Wyatt out of harm's way. It was a nice scene of my four-year-old playing in the flowers, but the gang is on the move, and we have to go.

A RECIPE FOR DISASTER

"Hello, Dave, my name is Randolph Beasley. I'm one of your neighbors." A breath of fresh air comes through the phone as Randolph introduces himself. Like me, Randolph can't tolerate what the criminal gang is doing to our community. "I'm running for office against Janice Rutherford," he says.

Until now, no one was opposing Janice in the coming election, and the deadline to run against her has passed. But when Randolph learned of corruption at the school bus stop, he filed special forms to get on the ballot against Janice Rutherford.

"I'm making a documentary," I say. "Will you talk about this on camera?"

"Sure."

Gianny drives up from Santa Barbara and places his movie camera on a heavy tripod in my dining room. Randolph sits down in front of my picture window overlooking the San Bernardino National Forest. Golden light shimmers through oak tree branches onto Randolph's weathered features. Randolph has dusted fingerprints and studied blood spray patterns in his role as a crime scene investigator — a CSI — and that's why he's here. Tears form in his eyes when he talks about the criminals who have taken over the school bus stop. At first I'm shocked that a law enforcement veteran would show such emotion, but then I feel relieved, because I finally found someone who believes as strongly as I do that the bad guys must be stopped before they kill one of the kids.

Finally, someone is taking a stand against Janice Rutherford other than me. I admire Randolph's courage to leave behind a safe, private life to run for office. Randolph says, "I don't have a snowball's chance to win this election. Janice has three hundred thousand dollars in her re-election fund. I put two thousand dollars on my credit card." Borrowing money to enter a no-win scenario? Now I *really* like Randolph.

> "It will take a miracle..."
>
> — Randolph Beasley

Sometimes I feel like the crazy guy standing alone in waist-deep sewage weathering a hurricane of corruption. But today, I'm not alone.

As a CSI, Randolph studies blood spray patterns from violent gangs like the one at Storybook Inn. He calls what Janice Rutherford is doing to our community a "Recipe for Disaster." I know we're in danger, but Randolph has investigated several hundred murders in his law enforcement career. He studies what criminals do to their victims, and he says there have already been murder cases related to the lawlessness at Storybook Inn.

"I've investigated crimes against children—" Randolph's eyes fill with tears. His voice trembles, and he suddenly seems like he doesn't want to finish his sentence. I don't blame Randolph for getting upset. Any human being with a soul recoils when they learn about the hundreds of criminal elements piling into our school bus stop and lighting cigarettes. Thank God Randolph Beasley is running against Janice Rutherford.

BEWARE OF DOG

In *Book One*, I told the story of a vicious dog I rescued off the street. It took me three hours to get a leash on the crazy beast. She was running loose in a bad part of town. At the time, I didn't have a clue how I was going to take care of a dog, but I put a leash on her and named her Switch.

Somehow, Switch and I got through it, one day at a time. Now, ten years later, Switch the Angry Dog growls and bares her teeth as Storybook gangsters shout profanity and lurk through the forest behind the hotel. As the hoodlums get closer, Switch issues a series of swift warning barks to alert me of the approaching threat. She postures up and positions herself between me and the thugs. She carefully sniffs the wind the way a police officer would check IDs in a traffic stop. I can't smell a thing, but to Switch, the air carries a signal from each of the bad guys. A low growl thunders through her chest and rumbles through clenched K-9 teeth.

Sky Forest used to be the home of Santa's Village, but the Christmas theme park has been padlocked by crooked county officials. And now the sleepy little town has become a hive of criminal activity. The bad guys have me outnumbered and surrounded, but if they want to get to me and Wyatt, they'll have to go through the fiercest protector I have ever known. I wouldn't have this peace of mind if I hadn't spent ten years scrambling to take care of this crazy dog. Now the crazy dog is taking care of me.

ESCAPE FROM STORYBOOK INN

WHITE, PUFFY CLOUDS SWIRL FOUR thousand feet above the people of Los Angeles. But my house sits on a mountain so tall that the house is in the middle of those clouds. As I walk through the grass, my boots are soaking wet as cloud mist sticks to my feet.

Sky Falling
A Documentary

But in this magical realm of clouds, a storm is brewing. It's not a rain storm or a snow storm. It's a perfect storm of human evil.

As a school bus rolls by, a gang of shadowy figures fly out of the Storybook Inn. They run across my yard the way lions hunt a gazelle.

"San Bernardino County Sheriffs," a dispatcher says.

"There are gangsters running through my yard."

"Who are they?"

"I don't know. They came from Storybook Inn."

An hour goes by with hooded figures darting around the hotel shouting profanity. But sheriff cruisers do not come to our rescue. Another kind of county vehicle shows up right on time as a school bus picks up children in front of Kory's office.

Hotel van drivers rev engines as they swerve around the school bus in pursuit of escaping guests.

Children walk amongst gangsters who are jumping in and out of those dreaded vans. This is the craziest situation I've ever seen. Most people in my town are afraid to speak of it publicly. I can honestly tell you that if it weren't for this book and our documentary, you would never know what happened at our school bus stop on October 9, 2013.

As darkness falls, the "don't fuck with us" gang hunts their prey. Their flashlight beams make the hotel look like a Nazi prison camp. You don't want to come between these Nazis and their money. But loud voices in the forest tell me someone has...

Sky Falling
A Documentary

Back in my house, Switch the Angry Dog growls and makes those little half-barks that wolves use to warn the pack of danger approaching. She peers through the window glass into the cloud-covered forest where shadowy human figures dart between the trees. She half-barks again and looks back at me. I slide twelve-gauge shotgun shells into the seven-round magazine of my Winchester Defender as a storm of bad guys swirl through the woods around Storybook Inn. It's going to be one of those nights.

THE THIRD WORLD

"Dave, I can't be seen talking to you," one of my neighbors worries out loud.

"Oh? Why is that?"

"I'm afraid they'll climb our fence and murder us." My neighbor explains his fear of the "don't fuck with us" gang and why he stopped answering my calls. I also notice he stopped posting comments on Facebook.com/SaveSkyForest.

This is supposed to be the United States of America, a free country ruled by laws. But lawlessness is causing good people to fear retribution from Storybook thugs.

"Dave, I can't talk to you anymore," says another neighbor. "They've already attacked my business. I'm sorry." She looks down at her shoes. I suppose she might feel embarrassed for not sticking up for the kids at the school bus stop. That's the last I heard from her. She never calls again to ask for an update or to offer information about the school bus stop like she used to.

"Dave, you can't mention me in your book!" pleads the owner of a popular tourist business in Sky Forest. That's too bad. Innocent victims are pouring into the Storybook Inn, partly because no one is willing to speak out about the danger. Innocent victims enter a hotel that is advertising nationwide for vulnerable customers who bring fat insurance checks. They have no idea they're entering an illegal, unlicensed business. The rich payout money for each hotel guest is worth more to gangsters than the life of the guest. But eyewitnesses are too afraid of the criminals to warn the traveling public.

After living this for four years, I can honestly say that people are right to be afraid to speak out against the "don't fuck with us" gang. I've seen what can happen if you stand up to crooked cops and corrupt politicians. A guy can get his truck impounded. A guy can be killed in a fiery car crash that's made to look like an accident. And no one will know.

Now I'm reminded of my time in Peru. In *Book One*, I told the story of how I got caught in a whirlwind of third world corruption. I had two options, life in prison, or death by murderous thugs. I called the American Embassy for help, but I was told that in Peru, my life was worth less than my belongings. If you read my first book, you know I was lucky to escape Peru with my life. I was so happy to get out of the Third World and return to the safety of the United States. But now in Sky Forest, I fear the same murderous corruption that almost killed me in Peru.

In Peru, I was saved by a guardian angel named Maria Reiche, who pushed the thugs and corrupt officials back so I could flee the country. But here in Sky Forest, no one is coming to help me. In fact, it's the opposite. People are going out of their way to tell me that I'm alone in this fight against a gang that outnumbers me a hundred to one.

Every night at Cloud Nine, I get on my hands and knees. I kick off my shoes and pray like a Muslim with my forehead against the hardwood floor. "Dear God, please send some rich, powerful person to fight for us. Please protect my neighbors and give us our beautiful town back."

You're probably sick of that prayer. I know. I'm sick of my life right now.

FOR SOME, IT IS ALREADY TOO LATE...

"Dave, someone went missing from Storybook Inn," says a neighbor.

Another missing person? Uh oh. We better document this.

"Can I interview you for the documentary?" I ask.

"No way. They'll ruin our business," says my frightened neighbor.

"OK. Got it. Thanks." Even with victims disappearing from Storybook Inn, people are afraid to be seen talking about the place. My neighbor won't tell what they know on camera. I used to take it personally when people refused to help the cause or warn people of the danger. But I've grown up a bit and learned to accept it. It's a bitter pill. But I never ask twice. At least my neighbor cared enough to give me the tip.

He continues, "The victim's family is posting homemade "Missing" posters all over town. You might want to check it out."

"OK. Thanks. I'll check it out."

Right away, I can see something isn't right with the latest disappearance from Storybook Inn. First of all, if someone is missing from the hotel, why aren't hotel workers searching for them? Why aren't Storybook people putting up the posters, asking around town, and helping to find the missing person? Don't they want their missing hotel guest to be found?

I put a copy of the missing person poster on Facebook/SaveSkyForest.

I'm told that by posting on Facebook, I did more than the entire staff at Storybook Inn did to help the victim's family find their lost loved one,

The missing guest is a twenty-one-year-old man. He has a great smile in the picture on his "Missing" poster. But with a gang of criminals running that hotel, I hope the kid makes it home to his family in one piece.

Save Sky Forest
November 1, 2013

Please help find this troubled person, missing from the drug rehab. Our prayers go out to him and his family. If you have info, please contact the Sheriff: 909-336-0600

NATHAN CRAWFORD

NATHAN CRAWFORD SCRATCHES HIS CHIN and stares intently at his computer screen.

"Is this the hotel?"

"Yeah. That's the hotel." I point a half inch to the left. "Here's the school bus stop."

I point a half inch down. "That's my house."

"OK," he says. "And another guest is missing?"

"Yes. That's the latest," I say.

I'm lucky to have a friend like Nathan. He organized a team of scientists to build a 3-D image of Sky Forest. It's amazing. You can fly into an actual image of Storybook Inn and look for evidence of crimes around the hotel. We can also look on his map for clues to the disappearance of the missing hotel guest. This will be a Godsend to help us keep an eye on the crazy hotel and could be used to help families of people who disappear from Storybook Inn.

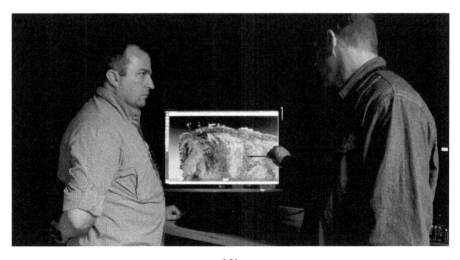

Nathan stares at his 3D map of Storybook Inn for a long minute. I think I can see the wheels turning in his genius mind. He taps his forehead as if the answer has just popped in and says something astonishing, "Dave, you should get a press credential."

"What? I'm not a reporter," I say. Nathan is so much smarter than me that his ideas sometimes make no sense the first time I hear them.

He explains, "Every American has a right to ask questions of corrupt politicians. You can get a press credential. I can show you how."

As usual, I never thought about it the way Nathan sees it. But Nathan explains civilian applications of his military surveillance technology. If you've seen the movie *Déjà Vu*, where FBI agents use technology to go back in time and see what happened at a crime scene, then you have some idea of what Nathan's team can do.

Nathan shows me a press credential and shows me how to use it. "Just wear this around your neck, Dave, and you can go anywhere and ask anyone any questions you want, even corrupt politicians. They can say anything they want to you, but they have to give you access. And then you can document the answers."

Nathan is already thinking outside the box I'm in. He's a good friend who takes note if I don't take his advice. So I immediately take the steps to get my press credential just like he showed me.

If you're trying to get away with murder at Storybook Inn, you've got a new problem...

Arnold, David
Correspondent

Gender: M
Height: 6'1"
Appx. Weight: 185
Hair Color: Brown
Eye Color: Hazel
DOB: 2/27/72

ISSUED:
12/29/2014

CARD EXPIRES:
12/31/2016

ID# 7292314

It feels so good to put on my press credential and walk up to Storybook Inn with my camera. They can tell me to get lost or go fuck myself like usual. They can even call crooked cops to get rid of me. But now I'm a credentialed member of the media. It's not that simple.

Right away, I start getting answers to questions at Storybook Inn. The gang-bangers seem very uncomfortable when they see the word "Press" hanging around my neck. Guys who used to menace me physically back away when they see the media badge.

A MESSAGE FROM STORYBOOK INN

"Whoa!" yells my helicopter pilot. He clings desperately to his flight controls. The Alaskan winds of *Deadliest Catch* toss our helicopter like a child's toy. I'm trying to focus my camera for Discovery Channel, but the whole aircraft shudders from the violent wind gusts. I grab my Cineflex laptop controller to keep it from getting thrown off my lap.

"I'm full-down-collective and we're still climbing!" my pilot shouts. I look around. Everything I see is trying to kill us, from the spiraling winds to the freezing Bering Sea waves. It's like trying to fly a helicopter through a giant dishwasher.

Right now, my pilot has a problem, because he's depressed his flight controls full down, but the murderous winds are kicking us up like a soccer ball. The pilot can't control the machine. But fortunately, he's a battle-tested Alaskan pilot who doesn't panic; instead, he lets the torrent push us up. Then he carefully musters power in reserve for what he knows will come next. *Wham.* Like a windy sledgehammer, the storm smashes us down. As we tumble toward the Bering Sea, my pilot doesn't panic. He pulls gently on the collective lever. A needle on his instrument panel rises until it's right against the redline. That's it. If he pulls with more power, the helicopter will come apart. But after using all our power, the little helicopter is still plummeting in the deadly downdraft. The state of Alaska is littered with the corpses of aircraft just like ours that succumbed to a murderous storm just like this one. The Bering

Sea storm waves grow large in our windscreen as we careen in the storm. But our pilot masterfully pulls and pushes all four control levers in unison to nurse the little flying machine back up and away from the storm waves. I look at the crab boat in my Cineflex viewfinder. The white-capped waves around the boat suddenly flatten as the downdraft hits them. Then the waves rise in another direction as an air cannon of wind drives them left, then right, front, then back. It is a full-throttle spectacle of deadly force.

"That's it. We're done. I'm sorry guys. We can't fly anymore." My pilot pushes his cyclic stick toward Dutch Harbor, and we shake and shudder with our tail between our legs. I look over the pilot's shoulder to his power gauge. He's pulling all the way to the mechanical limit just to keep us alive in the weather *Deadliest Catch* is famous for.

When I land back in California, I breathe a sigh of relief. I survived another storm season on *Deadliest Catch*. But as I crank up the diesel, I wonder what I'll find back at my house. I wonder if everyone is still in one piece at the school bus stop.

The lethal elements of Alaska are not for the faint of heart. But since my town was taken over by organized crime, I no longer feel safe in Sky Forest, either. Every night when I close my eyes, I hope we all make it to see the morning. But with the gang running amuck and law enforcement turning a blind eye, I never know.

My iPhone flashes. The screen says I have a missed call and a voice message…from the Storybook Inn. What's this? I press the play button. "Hi Dave, my name is Cyndi Doyle. We want to talk to you about Above it All Treatment Center. My son went missing from Storybook Inn six months ago…"

This is a dangerous time. Criminals are making hundreds of millions of dollars at Storybook Inn. Who knows what will happen when they find out a hotel guest reached out to the last man standing against them? But I have a strong feeling someone has to help Cyndi and her family. So I call her back and invite her to my home.

In the kitchen at Cloud Nine, I mix bananas and whole-wheat flower into pancakes for Wyatt. I always keep Wyatt away from Storybook people—but not today. When the knock comes on our door, I swing it open for Cyndi and her husband Shannon. As I look at the grief-stricken couple, I hear a voice of someone I can't see. The voice says, "Treat them like family."

What in the hell? That's easier said than done. These are victims of the "don't fuck with us" gang. Treating them like family can get me and Wyatt in trouble. But on the other hand, I haven't let good judgement stop me before. So I listen to the voice and open my home to the Doyle family.

"C'mon in," I say. "Sit down. Can I get you something to drink?" The Doyles look they've been run over by a train. I suspect dealing with the gang at Storybook Inn would break the will of most people.

They can see Wyatt through the glass doors where he's

watching a Disney movie in front of the fireplace. The Doyles lost their son to the "don't fuck with us" gang. I'm trying to keep my son safe from the same gang.

Remember the kid on the missing poster? Cyndi and Shannon say that's their son, Donavan Doyle. No one has seen or heard from him since he checked into Storybook Inn.

I think about their predicament. Maybe if people learn about Donavan's disappearance, it will help others to reconsider before sending loved ones to Storybook Inn. So I say, "I'm making a documentary about the hotel. Can I use your story in the film?"

Appearing in the documentary and talking about Storybook Inn is a dangerous choice for the Doyle family. The gang can advertise nationwide for addicts like Donavan so long as no one knows that criminals are running an illegal rehab at a school bus stop. People who speak out and call attention to Storybook Inn and missing hotel guests can be targeted for retribution.

I've talked to people who've been inside the hotel. But they will not speak publicly or show their faces in the documentary for fear of the gang and crooked politicians. Think about that. People are disappearing, and witnesses are afraid to say a word about it.

Charles Ricciardi's words echo in my head, "Make sure everybody signs a release." So I hand the Doyles a contract they're happy to sign, obligating them to appear in the documentary in return for help finding their son.

I turn on my little camera and offer Cyndi Doyle a seat next to my big picture window. The soft light of the forest illuminates her weary features. Cyndi takes a trembling breath and tells the story of her son's disappearance. "The day that he went missing, he was in an altercation with one of the counselors." Cyndi seems shocked by this, but she doesn't know the gang at Storybook Inn like I do. I know they sometimes fight with the patients. But Cyndi says worriedly, "I don't believe that is the way a trained professional treats a patient." I feel bad for Cyndi, but I know what she means. I've knocked on the door at Storybook Inn. I've never seen a "trained professional" answer the door.

Cyndi says Donavan ran away from Storybook Inn, trying to escape. I remember the night. Storybook thugs ran around my house hunting Donavan with flashlights. Cyndi says Donavan hasn't been seen or heard from again.

SKY FALLING
A Documentary

We called the sheriffs when the gang chased Donavan through my yard. I didn't know who Donavan was at the time. I only knew the hotel was hunting an escaping guest.

I'm trying to be strong and supportive, but I get goosebumps when Cyndi describes the way Storybook thugs and County officials hid the truth about what happened to Donavan Doyle. Tears pour down Cyndi's face as she says that even news media appear to be hiding the story of Donavan's disappearance.

SKY FALLING
A Documentary

It's a strange situation. Millions of dollars are pouring into the "hotel." That kind of money can pay for a massive search effort. But Cyndi and Shannon say no one has looked for their son, who disappeared six months ago. She says the sheriffs appear to be working with the gang at Storybook Inn to cover it up.

Los Angeles is one of the great media capitals. But Cyndi's tears turn bitter when she complains that "no story was ever run." It makes me angry, too. The media is supposed to be a watchdog for people like us. But after initial media reports about the school bus stop, I can't get the media to return my calls. I agree with Cyndi. The media is burying the story. It appears powerful members of the media and law enforcement know the story of Donavan Doyle but are helping the bad guys at Storybook Inn keep it a secret.

There are no rehab workers here to help the Doyle family, even though they paid a king's ransom for drug counseling at Storybook Inn. The sheriffs aren't here to help solve Donavan's disappearance. All the Doyles have are two family friends who paid for the trip to Sky Forest. But no one else is helping look for their missing son. Cyndi's voice rises to the peak level of my audio meter when she weeps over her son's disappearance. It is a heart-breaking sound. Tears roll down her face. "I just want to bring my son home!!!"

We called the sheriffs when Storybook workers chased Donavan through my yard. But sheriffs maintain that they've never had a single call to Storybook Inn. I guess we're supposed to believe that nothing happened to Donavan Doyle that night.

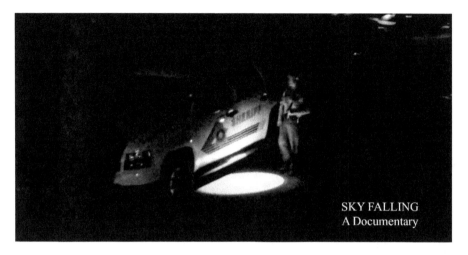

SKY FALLING
A Documentary

I have to turn the camera off because, while Cyndi is talking, Wyatt is jumping up and down on the couch behind her. My camera can see Wyatt through the glass doors. I hate to interrupt Donavan's mom, but I suppose it's important to note that my son is in the next room while she's talking about what happened to her son. The gang that took Donavan lives next door to us. Tonight, Wyatt and I will have to sleep in the shadow of that wretched hotel.

Cyndi and Shannon are convinced that Storybook thugs did something to their son and are lying about his disappearance. I can tell they fear the worst, that the "don't fuck with us" gang murdered their son, Donavan Doyle.

Next, they say something even more shocking. They believe the San Bernardino County Sheriffs know what happened to Donavan and are hiding it. I place a nervous hand on my little camera screen and make sure I'm getting the sound. If the Doyles are right, none of us are safe from Donavan's fate. Anyone who investigates the "hotel" may be the next to disappear. As I think about the scary situation, that voice comes into my head again, "Make them feel loved."

Make them feel loved? What in the hell am I supposed to do? This isn't safe. I have an organized crime ring so big, they actually outnumber the residents of my town.

I am not prone to quick thinking, but suddenly I'm crystal clear on what to do next. It's like that time a voice told me to put my comatose mother on a telephone call. It makes no sense, but I know exactly what to do. So I look at the terrified parents in front of me and say, "I can do two things for you right away. First, we'll hold a candlelight prayer vigil for Donavan. Second, I'll walk from where Donavan disappeared, down the mountain, all the way to the city."

I have no idea how to get from here to the city. There are twelve miles of vertical mountain and rattlesnakes between Storybook Inn and San Bernardino. But I'm overwhelmed with a sense of absolute purpose. I know this herculean effort will let Cyndi and Shannon know that their son is important. Besides, not having a plan never stopped me before.

The voice chimes in like a bell, "Treat them like family." OK. I'll do it.

"You guys always have a place to stay in Sky Forest," I say. "Any time you want to come and look for Donavan. You can stay at my house."

Cyndi and Shannon look like they've seen a ghost. No one at

Storybook Inn has helped them in six months. And in one meeting, I've opened my home, offered my resources, and I'm starting a one-man crusade to find their son.

"DON'T DO IT DAVE. THEY'LL KILL YOU..."

Last year when I announced my plan to check into Storybook Inn and stay overnight, my neighbors panicked. "Don't do it Dave; they'll kill you!" Now their words echo in my head as I plan a bigger trip further inside the hotel—this time to find one of the missing guests. Doyle family friend Larry Pasquale shakes his head in disbelief. "I can't believe you're doing this. I wouldn't do it."

Larry has experience with gangsters like those at Storybook Inn, and he says there's a serious danger in crossing them. Larry is right. It's dangerous to fuck with the "don't fuck with us" gang. But I heard a voice in my head. The voice said to "Treat them like family." So now I'm on a mission. Damn the torpedoes.

I thumb through my calendar, and I see a special marking on one of the days—Memorial Day. That's it! That's the day I'll walk over to Storybook Inn and search for Donavan Doyle.

We'll have a candlelight prayer vigil the night before. I know that will make Cyndi and Shannon feel loved. I can't undo what the gang did to their son, but I can treat them like family. So here we go.

I'M DOING IT ANYWAY

IN THE HISTORY OF HAIR-BRAINED schemes, this one is a doozy. I'm going to hold a public gathering at a street gang headquarters so we can light candles and pray for someone they might have murdered. If there's a better way to bring trouble, I can't think of it. So naturally, I invite my friends.

A hair-brained search won't really do the Doyles' much good. But if I could draw a crowd for their son, that would be something. So I ask my friends to give up a holiday weekend to come and help with the search. Storybook workers claim there's no sign of Donavan at Storybook Inn. But I know Donavan's family will feel loved if people come out to Storybook Inn and make an effort to find him. At least we can cross some places off the list of where Donavan could have gone when he disappeared from the hotel.

Nowadays, Memorial Day is just a long weekend for beer and BBQ, but I don't drink beer. And I like the idea of setting fun aside to memorialize someone who is lost, but I'm not sure what my BBQ-eating, beer-drinking friends will say. I grab my phone and call Gianny. "I want to do a prayer vigil and wilderness search for that missing kid from Storybook Inn."

"OK. You want me to bring my Red Camera?"

What? I was expecting to hear an excuse for why he can't make it. Who has time for a cockamamie search when there's BBQ and beer to be had? I'm surprised he said yes to that. But Gianny didn't even take a minute to think about it. He's on board.

"Ummm…OK. Yes. Please bring your camera," I say.

I shouldn't ask Gianny to climb a mountain with his heavy Red camera. I have no money, and that camera is too expensive to drag through the dirt. I have no reason to believe we'll find anything at Storybook Inn. Gianny should just use my little cheap camera. But before I can explain, he says, "No problemo amigo. I'll be there."

Gianny is a nice guy, but this is ridiculous.

Gianny also agrees to film the prayer vigil. I think it will mean the world to Donavan's family to see people like Gianny going out of their way to show love and support, without judgement, at the prayer vigil. And the sight of Gianny's heavy-duty camera is a sign that something important happened on October 9, 2013. Donavan is more than a notch in the belt of criminal hotel workers. He's a human being, and his disappearance is worth looking into.

I extend an invitation to Storybook workers to come to the prayer vigil and wilderness search. This will be a great opportunity for anyone with kindness in their heart to show support for Donavan's family. I know I'm guilty of pie-in-the-sky thinking, but I've been praying for gang-members to fight their violent urges and show kindness to their patients. Maybe one of them will have a little sympathy for Donavan's loved ones. This will be an opportunity for them to set their "secret business" aside and give me a hand.

I'd love some volunteers to help search the wilderness, but someone could get hurt. In fact, if *I'm* planning this thing, someone is probably going to get hurt. So I call Hugh Campbell (the one who told me I don't know what I'm doing). Hugh knows better than to get involved with my hair-brained scheme at Storybook Inn. This might be an awkward phone call. But I'm thinking about Hugh's mountaineering experience, so I dial his number.

"Hello?" asks Hugh.

"Hi Hugh! I'm planning a prayer vigil and volunteer search for a missing kid at Storybook Inn. I don't really know how to organize this. Do you think you could lead the search expedition for me?"

Hugh is one of the smartest people I know. As a retired aerospace engineer, he can calculate every risk and reason not to do this. But without hesitation, he says, "Sure. I can do that."

I'm shocked. I was expecting some sort of argument from Hugh. But he didn't even hesitate.

"Thank you!" I say with a huge sigh of relief.

One of the things that the Doyles said, with tears in their eyes, is, "The sheriffs are lying. They aren't helping." So I'm thinking about Randolph, who's a decorated CSI, who sometimes appears on HLN to talk about unsolved crimes. Cyndi and Shannon will be blown away if I could get a renowned forensic scientist to search for evidence of Donavan's case. So I grab my phone and dial Randolph's number. "Randolph, will you come out and search for evidence of a kid who disappeared from Storybook Inn?"

I'm really out of line now. You're supposed to call a crime scene investigator when you have a crime scene. But I don't have anything. In fact, I have zero information to say anything happened to Donavan, except for Storybook Inn workers who claim he just wandered away under his own power. Instead of a crime scene, I have a cockamamie plan to wander the woods where the Storybook gang says we won't find anything. I can honestly say that Randolph doesn't have time for this. But I know it will make Cyndi and Shannon feel loved. So I ask anyway.

On the other hand, there's no way Randolph can say yes. He's in the final days of his Hail-Mary campaign against Janice Rutherford. He needs every second of fund-raising and vote gathering. But without hesitation, Randolph says, "Yes. I'll be there."

I'm floored. Just like that, and without a word of question or argument, I have a crime scene investigator coming to Storybook Inn.

But I'm not done making unreasonable requests. I want a helicopter. After being jerked around by Storybook gangsters and the sheriffs, I know it will mean the world to Cyndi and Shannon to look up and see a helicopter thundering through the sky for their son. Just the sound of helicopter blades will make them feel loved. So I call my long-time friend and helicopter pilot Craig Dyer. Craig is a veteran of Desert Storm, where he flew U.S. Army Blackhawks. Now Craig has a busy personal and professional life—always building a house or doing military stuff. There's no way he can agree to fly a helicopter for our crazy wilderness search, which will force him to give up his holiday weekend and cancel family plans. But for some reason, I don't care about all the reasons not to do it. I just pick up my phone, dial Craig's number, and say, "Craig, I need to find a pilot to come and help search the forest for someone who's missing." To my surprise, he says, "Yes." And he didn't hesitate. Just like that, we have a helicopter pilot.

Randolph says he has the phone number of a woman who suffered a terrible loss when her son died at a rehab, so I give her a call. Wendy McEntyre doesn't know me. She's never heard of the Doyle family. But I ask her to come out on Memorial Day and take care of Donavan's mom and dad for me. I know it will mean the world to them to have another parent who understands what they're going through. There're a million reasons for Wendy to say no to this. For starters, no one at Storybook Inn said Donavan could be in the woods behind the hotel. We don't even know what we're going to find on this search. But without hesitation, Wendy says, "Yes."

Thank God Wendy's coming. I'm worried about Donavan's parents, and I won't be able to comfort them while climbing the mountain on the crazy search. Sight unseen and no questions asked, Wendy agrees to come out and take care of Donavan's mom and dad for me.

Just like that, without proper cause or justification, my friends and some strangers have raised a pea-brained, amateur scramble into a serious search expedition with experts in every field. It was a lot to ask, but I got a ground search leader, a crime scene investigator, a helicopter pilot, and a victim counselor.

NEEDLE IN A HAYSTACK

"Above it All Treatment Center is not offering any help whatsoever." — Renae Pasquale, Doyle Family Friend

CYNDI DOYLE WIPES TEARS FROM her eyes. She takes a trembling breath and says, "We got no support from Above it All. We got no support from Twin Peaks (Sheriff) Station." Cyndi says Sheriff Detective Camacho failed to deliver a report on Donavan's disappearance. I get chills when Cyndi describes sheriffs "not filing reports," "not investigating," and "tipping off bad guys at Storybook Inn."

Something stinks. And it's more than just the sewage that has been overflowing from the hotel. If Donavan is a needle, the world is a haystack. Even worse, the Doyles think the sheriffs are helping make their son disappear. As unthinkable as that is, I see their point.

Outlaws at an illegal clinic fought with a patient who was never seen again. Now the outlaws claim they have no idea where Donavan went, even though they were seen chasing him when he disappeared. And no one has seen or heard from Donavan since.

The sheriffs claim there's nothing to report, even though they were called to Storybook Inn when Donavan was fighting to escape the hotel. The lies and cover-up of Donavan's disappearance tell us Donavan could be anywhere and nowhere. It's as if he never existed. That can't be. We're talking about a human being. Human beings don't just disappear, unless they're made to.

And that's the key. It's the reason my neighbors are so afraid of that hotel. Everybody knows that any of us could be in Donavan's place, where our life is worth less than the millions of dollars flowing to Storybook Inn.

I've been warning the public that the illegal rehab is a danger and should be shut down, but I could be made to disappear just like Donavan. Donavan's disappearance has been swept under the rug at Storybook Inn. The sign out front is still advertising "Lodging and Dining" to unsuspecting people just like the Doyles.

So where do we begin? How I find somebody that the bad guys don't want found? I don't know. But I'll hand this over to the good guys who are volunteering to help on Memorial Day. I ask our helicopter pilot, ground search leader, and CSI to have a conference call with Donavan's parents to map out a strategy for Memorial Day. Craig, the helicopter pilot, does most of the talking.

He opens with a series of questions for Donavan's mom and dad.

"What was Donavan's condition when he disappeared from Storybook Inn?"

"What medication was he on?"

"Did he take his meds that day?"

"Does Donavan have any disabilities?"

"How did he sound on the phone before he disappeared?"

"What was his state of mind when you last spoke to him?"

"I need to know, when facing vertical terrain, would Donavan go up or down?"

"Does Donavan have mountain-climbing experience?"

Craig's expertise as an army aviator comes into play here.

Donavan's mom looks shocked and a little relieved. Her eyes grow wide, and she says, "That helicopter pilot asked me more questions in one phone call than the sheriffs did in seven months!"

I'm not surprised. I've seen how Craig handles a mission. He pours over terrain maps and teaches his son Sam how to plan the helicopter search mission. "OK. On Memorial Day we're going to fly to Sky Forest and establish radio comms with Dave Arnold." Craig points to a satellite image of Sky Forest and says, "This is Dave's house."

It's hard for Sam to see anything on his dad's map, but to a terrain expert like Craig, the little lines and bumps read like a novel. "That is the hotel. This is the area we want to concentrate on."

How can Craig be so sure we'll find something at the hotel? Not a single hotel worker pointed to Storybook Inn. But Craig is a genius of planning and strategy. The bad guys at Storybook Inn will have a hard time hiding crimes from Craig and his helicopter.

Craig says he has a homework assignment for me. "Dave, can you climb into the canyon behind the hotel? Before we fly over this place, I want to know if a person like Donavan could get through there on foot."

I never would have thought of that. But Craig's military mind is attacking the secrets of Storybook Inn like a special forces mission.

Craig wants me to climb into the backyard of an organized crime ring and poke around? That's ridiculous and dangerous. So of course I say, "Yes." And I call Hugh. "Will you climb behind Storybook Inn with me? Our helicopter pilot wants a terrain check before Memorial Day." I must be on a roll, because Hugh doesn't hesitate. He usually stops me when I'm starting down a hair-brained path. Hugh has every reason not to do this. Remember that scene in *The Godfather* when the next-door neighbors poke around in the bushes behind Don Corleone's house looking for dead bodies? No? That's because the mafia would kill the neighbors and bury them next to the other bodies. Poking around the Storybook Inn will be considered fucking with the "don't fuck with us" gang. This could end badly. But God bless Hugh. He says, "Sure Dave, we can do that..."

IN HARM'S WAY

"WHOA! I DON'T LIKE THIS!" My Alaskan helicopter pilot pulls hard on the collective. I sink in my seat as the Long Ranger helicopter's engine screams in response. He pulls the aircraft into a hard turn to dodge the volcano erupting below us. "We're out of here!" he yells as his survival instincts turn us away from the bubbling volcano.

My pilot is right to be afraid. Last night this volcano blew an eighteen-thousand-foot mushroom cloud of poison gas and island parts. Most of the island was vaporized.

Today we're filming for volcano scientists who are trying to improve warning systems that could save victims from eruptions like the one below our skids.

Airlines have been warned not to go near this volcano. The FAA put a no-fly-zone around the island. So until we flew into the volcano, no one knew what it looked like. We're capturing the first human look inside an erupting volcano with our GSS Red camera.

At first my pilot was excited to fly over the volcano and witness the most awesome power in nature. But now he's panicked by the power that's bubbling toward my camera. But I'm just annoyed. "Hey! Hold on." Now that we're here, I want to film this thing.

The pilot nervously halts his turn and swings the control levers to hover back toward the steaming volcano. The ocean ripples to about a 7.0 on the Richter scale as the volcano bubbles below our skids. But the GSS holds our Red Cinema camera perfectly still so I can capture images of the eruption. I'm sure volcanologists will enjoy these images, but the pilot is right. We're in mortal danger, and we have no way to escape from the explosive power that obliterated the island. We should go.

But I completely lost sight of the danger, and now I only see the Red Cinema Histogram counting perfect 6K frames as the volcano bubbles up to meet our helicopter. No wonder Nathan said I'm thick as a brick. My pilot abandons his survival instincts and hovers above the volcano so I can continue filming. We're getting great volcano footage, but we could be killed at any second. You know what? Sometimes people *shouldn't* listen to me.

We finish filming the eruption. My pilot pulls on his cyclic and we fly fifty miles back to a distant island. We set our skids on the ground and finish one of a thousand missions I've flown in twenty-three years of cheating death.

I get out my wrenches and take the camera off our helicopter. I airline it back to LA and pick up Wyatt. Wyatt and I don't talk about helicopters or volcanos. We just climb into that big, noisy truck and sit in LA traffic. After two hours, we leave the cars and road-ragers behind and turn onto Rim of the World Highway. Graffiti and razor wire give way to meadows and oak trees as we make the fairytale climb to Sky Forest.

It's like driving through a surrealist painting as I steer our diesel truck into a cloud forest. I step out of the truck and take a deep breath of clean, mountain air. A whisper of wind soothes my ears.

"Dad, make a fire so we can be cozy." Wyatt sinks into the couch as the clouds wrap around our house, and I build a fire in the great stone fireplace.

As the flames light Wyatt's tiny features and the fire warms our toes, I put on a cartoon movie for Wyatt. I sip coffee and Wyatt has hot chocolate as Switch the Angry Dog guards the door.

Life is good.

As the movie credits roll, I pack Wyatt off to bed and settle back into my spot on the couch in front of the fireplace. Our home has a master bedroom suite, but I prefer to lay on this couch in front of the fire and watch the moon rise through the trees. That's how I fall asleep in Sky Forest.

But I'm awakened by movement. A four-year-old child has gotten out of bed and snuggles into my blankets like a panda-bear cub. It's ridiculously cute, but now I can't sleep. Outside our walls, predators from Storybook Inn are hunting in the dark.

I invited my friends and neighbors to the predators' lair. Thank God our volunteers agreed to help Donavan's family. But as with the Alaskan volcano, I don't know if I'm paying enough attention to the danger. I'll do my best to keep everyone safe, but gangsters are making a lot of money at Storybook Inn. I believe they would kill to keep that money flowing. And sometimes people shouldn't listen to me.

TOYS

"I HAVE SOMETHING TO SHOW YOU." Nathan enthusiastically points to a small plastic toy.

"It's a toy," I say with the enthusiasm of a brick.

"No. Take another look." Nathan drives his toy into the sky with a remote control. "This is game-changing technology!"

Nathan is an expert in airborne intelligence, but I have no idea what he's talking about, or why he's so excited about a toy. On the other hand, Nathan is always right. So I take his advice and buy a toy just like his to search for Donavan.

As I push the throttle lever on my remote control, the battery-powered blades sound like angry bees. The little machine zips into the sky above Storybook Inn. When I look at the footage, I can see Nathan is right, as usual. The toy holds my GoPro camera over the secret business and shows me places it would take days to climb to, if I could go there without getting murdered.

I immediately put the quadcopter to use for Craig's terrain analysis behind Storybook Inn. The toy glides between the treetops searching for any sign of Donavan. One man climbing in the forest below Storybook Inn is a hapless struggle, but one man with a drone and a camera can search a wide area of the shadowy woods. Nathan was right. He's always right.

Now Donavan's loved ones in Fresno can put the memory card from my quadcopter into their computer and search the forest with their own eyes. If there's any trace of Donavan in the woods behind Storybook Inn, we'll find it.

If you work at Storybook Inn and you're hiding something, you have a new problem.

It's a toy. But it works.

SKY FALLING
A Documentary

But there's one problem. Nathan skillfully flies the little quadcopter around his office, but it's hard to keep my footing behind Storybook Inn, and even harder to manage a flying toy. Most days I bring a pile of drone parts back to Cloud Nine. It takes hours to rebuild the little machine, and the parts aren't cheap.

I glue the little quad back together as best I can and leave a pile of broken parts in my shop as I leave Cloud Nine and head back into the forest. I place the little machine on the flattest spot I can find behind Storybook Inn. "Please don't crash today..." I push forward on the control stick, the batteries and little propellers struggle in the thin, mountain air. A chill runs up my spine as I steer the flying eye toward the Storybook Inn. There's an eerie vibe at this "hotel," and I'm hyperaware that bad guys are watching me and they will not be amused if they hear or see my little toy descending with a camera into their backyard. Uh oh. I have a problem. I'm pulling back on the stick, but the machine falls straight down into the backyard of the hotel. It's only been flying a few minutes. But in thin Sky Forest air, the teeny aircraft is already spent. Instead of climbing back up to me, the little machine spirals straight into a patch of dead trees directly behind the hotel.

In 2003 the woods behind Storybook Inn burned, so now I'm climbing through the twisted, blackened trees listening for the SOS beep that the quadcopter makes after crashing. Meanwhile, I hear angry voices on the Storybook balcony. I dare not look up at them. These gangsters get furious when they see anyone looking at them. Someone climbing into their backyard is likely to cause a full-blown riot. But I can't deal with them right now. I have to finish the scouting mission for Craig, and I lost the only tool I have in these bushes.

What kind of a bush has two-inch thorns? I don't know, but I have an endless tangle of them between me and the sound of my beeping quadcopter. What I find, eight feet down inside a thorn thicket, looks like a plastic airline disaster. The fuselage is smashed. The camera gimbal is gone, flung into the wilderness.

I have to push my whole body through an acre of thorns to find the camera gimbal, wires ripped, and camera lens ruined. Bar none, I have crashed every quadcopter, and it's getting expensive. My credit card takes another hit with each busted drone and camera—not good.

Oh well, at least I have the footage for Donavan's family. I'm sure it will give them some comfort and peace to search this part of the dead woods behind Storybook Inn. Maybe the footage will bring a clue to Donavan's disappearance.

SCOUT MISSION

Sky Forest is beautiful, but we find a mix of dead trees, stinging plants, and venomous snakes behind the hotel. I love nature, but as I walk through the backyard at Storybook Inn, there's a palpable darkness. I have to be so careful that Hugh and I aren't seen by the criminal gang. Otherwise, *we* could end up buried down here.

Hugh stops here and there to teach me about the plants. "Look here, Dave, these are wild strawberries. Those are stinging nettles. Don't touch them."

"Hugh," I say, "no one could get through here."

"I agree." Hugh says. He surveys the wicked slope and thorn bushes. "Nobody could get through this."

"Donavan couldn't get through here, especially if he was sick and fighting with gangsters."

We go all the way to Switzer Park in search of passable terrain, but we learn that it's impossible to get from Storybook Inn to the places Donavan was seen when he fled from the gang. Something is missing from the hotel story.

Suddenly I stop in my tracks. I have a sheer cliff on my left, and impassable brush on my right.

Hugh's background as an engineer pays off. He checks his computer-generated map and points to the cliff. "It's too dangerous to go here on foot. This area will have to be searched by the helicopter."

When we get back to Cloud Nine, Craig listens to our ground report over the phone. He looks at the terrain map thoughtfully and points to a ravine behind Storybook Inn. "If Donavan is out there, you'll find him here."

What? I don't know what Craig is talking about. No one said Donavan went down there. But I wonder if Craig knows something I don't. He must be seeing something I can't in his terrain maps. And I wouldn't bet against Craig. He's found missing people with his helicopter before. Anyway, we got what Craig wanted—a careful study of the terrain behind Storybook Inn.

MOTHER'S DAY

MY MOM DIED YEARS AGO, so I don't have plans on Mother's Day. But I hear a voice that says I should drive to Cloud Nine. I can hear the words in my head, but I can also feel them in my heart. It's hard to explain. The voice says, "You have to do something for my mom." Somehow I know that "My mom" is Cyndi Doyle.

I'm not planning to drive to Cloud Nine today. The trip doesn't fit into my work schedule. But I climb into my old truck and fire up the diesel. As I prepare to enter LA gridlock traffic, I text Donavan's mom and ask her to drop a pin on the map where Donavan was last seen the day he disappeared. My phone lights up with a message, and a map with a pin. I look carefully at the map. Cyndi's pin is on the school bus stop in Sky Forest. She says gangsters attacked Donavan there, and he was last seen running away from the gang on foot. I shift my truck into "Drive."

After an hour, I roll through the last of LA traffic and turn onto Rim of the World Highway. The turbo-diesel sings as I weave through the mountain "narrows," a vertical granite passage that looks like something out of an X-box driving game. As I look to my right, there's nothing below the truck but four thousand feet of air. Driving to Sky Forest is like flying a plane. After 10 minutes of high-G turns, I pull into in Sky Forest with no other purpose than to "do something for my mom."

I park the truck at Cloud Nine and make my way to the school bus stop. I have to crouch behind bushes so the Storybook thugs can't see me. The school bus stop is crawling with killers. Donavan's dad said a gang of Storybook hoodlums chased Donavan out of the school bus stop to the place where I'm standing.

I hear loud voices. "Watch out, motherfucker! You better check your shit!" The mood at the school bus stop is always on edge, like a biker bar when somebody's about to fight.

Storybook gangsters are ever-present in Sky Forest. I don't want to be caught searching for one of their missing customers. I could end up missing just like Donavan. So I sneak around the school bus stop, carefully keeping out of sight, and start a grid pattern search.

As I move further from the gang, I enter a beautiful forest meadow. Thick, green oak tree leaves flutter in the breeze. I take a deep breath. I love the forest, but somehow the forest is different today.

The birds are acting strangely. Normally they avoid humans completely. As I search for Donavan, I notice there's a bird sitting an arms-length away. But this bird doesn't fly away like they normally do. The bird tracks me with its eyes. It's surreal. I continue on and find another bird, and I get the same bizarre bird stare. The birds are normally very nervous and flee if I come within ten feet, but this bird calmly sits just a few feet away. I don't sense the normal, twitchy bird panic, just calm and tranquility. The beauty and peace of the birds today is a welcome relief from the murderous tension of the school bus stop on the other side of the bushes.

I scour the forest but find no trace of Donavan. However, it feels like someone is here with me in the forest. I feel a strong presence. It's not scary. Even though I'm yards away from the thugs who chased Donavan into these woods, I feel calm. This scene is one of the most surreal experiences of my life.

David Alan Arnold

As I finish my search, I know it will mean a great deal to Donavan's mom that someone cared enough to do something for her on Mother's Day. So I send her a picture of the forest, and a report from my search around the school bus stop.

David Arnold
To: Cyndi Doyle
This is what my yard looked like yesterday...

May 11, 2014 at 6:47 PM
Sent - iCloud

In weather similar to the day Donavan disappeared, I walked through the cloud forest and just sat down in the grass, watched the puffy clouds float through the trees and marveled at what my neighbor calls "Heaven on Earth". Little did I know I was sitting below the turnout where Donavan was seen walking into the forest.
Today as I searched the same area, I was surrounded by beautiful song birds. They didn't fly away as I approachedd, they just sat on the branches with their brilliant colors and sang sweet songs. This is a spiritual place. Sky Forest is beautiful, peaceful and comforting to me. Regardless of where Donavan went on Oct. 9th, he walked in the clouds and passed through one of the most beautiful places I have ever seen. Whatever stress and pain he was feeling, I hope the forest brought him some comfort and peace.
And I hope you find him soon.
-Dave

My email inbox rings with a message from Cyndi. She says my search helped her to get through a very painful Mother's Day.

It's good to know that she felt loved and a little bit of comfort, but I can't take the credit. The search wasn't my idea. It's like the phone meeting with my Grandma and my dying mom. I didn't think of it. The idea came from someone else. I can't see them, but I can hear them. I can feel their words in my heart like a drumbeat. I know that sounds weird, but fasten your seatbelt. Things are about to get even weirder.

ROBERT LARIVEE

SHANNON DOYLE IS SUFFERING A father's worst nightmare, with no clue to his son's disappearance. On the other hand, several father and son teams are volunteering to help search for Donavan. Chaplain Robert Larivee flips open the lid of a laptop computer and slowly zooms into a Google Earth view of Sky Forest. To me, the screen is just a satellite view, but to Robert, the lines and shadows tell a story. He points to a dark area on the map for his sixteen-year-old stepson, David Wemyss.

"OK. Here's the hotel. We want to search here…see this line?"

Wemyss leans in to see a faint diagonal line, barely perceptible, under his stepdad's finger.

Robert continues, "I think this line is an old logging road. We can use that to get through."

The dead forest behind Storybook Inn is a heckuva place to climb into, with no expectation of reward except helping Donavan's family. But Robert and Wemyss are giving up their holiday weekend to help us.

"Dave, I'm going to bring my radios so we can keep in touch when we spread out through here." He points to an area of dead trees on the map behind Storybook Inn. Robert is a civil air patrol chaplain. He brings training from his specialized career searching for downed aircraft. Thank God he's here. I can't imagine leading a volunteer search expedition into the wilderness behind Storybook Inn without people like Robert to guide us.

Robert's stepson is a cross-country endurance runner. Wemyss' athleticism and energy will be a Godsend for the tough mountain climb into the forest behind the hotel.

WING AND A PRAYER

"You know Dave, I can't see through the tree canopy. I won't be able to help you in the forest," Craig says on the phone. Craig and his son Sam will be in the helicopter on Memorial Day to search for Donavan Doyle. And I know Craig's right. There are hiding places behind Storybook that even a helicopter can't find. We need a miracle. That's why I want a prayer service before the search. But I'm not a minister. I can't lead a prayer service.

Fortunately, St. Richards Episcopal church is across the street from Storybook Inn. In the past, I've asked the minister to pray for the addicts at Storybook Inn, and she led her congregation in beautiful prayers for the workers and guests. So I'm pretty sure she'll help us out.

I grab my iPhone and dial the minister at the church. I say, "Can we have a prayer service in your parking lot for a kid who is missing from the hotel?"

"No," she's says flatly.

The minister's words blow through me like an ice-cold wind. You might expect an explanation or words of comfort from a minister. But I get nothing…just awkward silence on the other end of the phone. I never imagined a member of the clergy would say no to prayers for a family in need. But she flatly refuses. In fact, she offers no help whatsoever. I'm a bit hurt by this; I only asked to use the parking lot.

Since I can't use the church for the prayer service, I ask a local businesswoman to use her event space down the street. She says, "Sure, Dave. You can have a prayer service here."

"Great! Thank you." I'm pumped. This new place is farther from Storybook Inn, but a nice, quiet gathering place for prayers. You might think that's the end of it, and I'm all set. But unfortunately, five minutes later, there's a knock at my door. I open the door and find the same businesswoman. Her head hangs low, as if burdened by the weight of bad news. She looks around nervously, as if she's afraid someone will hear, and says, "I'm sorry, Dave. We can't do it."

Another cold shockwave washes over me. Doesn't anyone believe in helping people? I politely ask, "Why? What is going on?"

"We can't be involved, Dave. They'll close our business down."

She explains that, although she already gave permission for the prayer service, she now realizes that the prayer service could be seen as a threat to the gangsters, and she's afraid to be seen helping the victims of Storybook Inn because corrupt county officials might close down her business as a form of punishment.

Again, I'm shocked. The local government is so corrupt that if you so much as pray in public for a victim of one of their scams, they'll attack you and ruin your life. Again, this seems to run exactly like the mafia. People are afraid to even be seen helping the victims.

"OK," I say. "I understand."

Now that we both agree there will be no prayer service on her property, she looks at me with trepidation. "You know they're making a LOT of money at that hotel?"

"I know."

She stares at me sadly, the way you might look at a condemned man on death row.

"Do you have a gun?"

"Yes. I've been sleeping with a loaded gun since this began."

Her eyes fill with worry as if she doesn't expect to see me again.

I politely close the door on the frightened business owner. This fight against Storybook Inn gets lonelier by the day. Now I have no place to hold the prayer vigil. But I already told Donavan's family it would be at the event space. Since the owner fears retribution, I'll have to tell the family it can't be done.

There's so much crime and corruption in Sky Forest that a church won't offer prayers. And others are afraid to be seen helping me, because everybody knows I'm standing against the gangsters. Some people are so afraid of the bad guys that they're careful not to be seen standing next to me at a prayer vigil for one of the victims.

Back to the drawing board. There's only one place I can think of to hold the prayer vigil. So I type up an email and let Cyndi and Shannon know that we'll gather at the last place their son was seen—the school bus stop. It will be loud because of traffic noise, but we'll pray on the side of the road because it's the only place we can use in Sky Forest.

Since the church won't help us, I ask Randolph to help me find a minister to lead the bus stop prayer vigil. He doesn't hesitate. "Pastor Mack will do the service."

I'm relieved. If you've ever been in a dire situation like this, you know how good it feels to have a member of the clergy come onto the scene. Pastor Mack to the rescue!

But I'm worried about the highway noise at the school bus stop. How will people be able to hear Pastor Mack praying above the car noise?

"Randolph, do you know anyone who has a PA system so we can hear the prayers?"

"I'll check."

"And we need music. Do you know anyone who can bring music to the service?"

Randolph make notes as if writing down a grocery list. He calls back a few minutes later. "OK, Dave. Our friend Steve will bring his guitar and his PA system."

I'm a little amazed that Pastor Mack and Steve agreed to help us. They don't know the victims of Storybook Inn like I do. They don't know the politics that caused everyone else I asked for help to say, "No." But thank God Steve and Pastor Mack are coming to help anyway. Now we can have music and prayers on the night before the search.

Randolph's wife Rita makes a flyer to advertise the prayer

service and call for volunteer searchers. God bless Rita and her friend, who sit at my dining room table stuffing envelopes with invitations and a personal letter from Donavan's mom. I buy stamps at the Sky Forest post office. And the nice lady at the post office places a flyer in every PO Box, including Storybook Inn. It doesn't take long to fill every box in town. Our post office is the size of a walk-in closet.

Without exception, everyone in Sky Forest is invited to pray with us and join in the volunteer search. I especially want the Storybook gang members who have pain or regret in the aftermath of Donavan's disappearance to come. Maybe they can find some peace by lighting a candle and praying with Donavan's family. Maybe one of them will tell us the one thing we don't know — where Donavan went when he disappeared from Storybook Inn.

FALLING

I'M HAVING A STRANGE DREAM that Tom Cruise is doing a movie stunt for *Mission Impossible*. But instead of stunt-men, he is using me and my Sky Forest neighbors for the stunt. Tom is animated. "OK. You guys jump off the cliff and turn this way!"

As Tom plans the stunt, I'm looking down the four-thousand-foot drop below the cliff. We're supposed to base-jump and spread out in separate directions while Tom Cruise falls with us. I can't figure this out. We're supposed to fly through the forest like wing-suited daredevils in a *Mission Impossible* movie stunt, except we don't have wingsuits. And we're not stuntmen. My neighbors haven't even been skydiving before.

Tom is pacing back and forth pre-visualizing the base jump. He's so enthusiastic and confident, but I can't see how this is going to work. A Billy Joel song is playing on a loud-speaker as we get ready to jump without parachutes, "Only the Good Die Young." I'm shaking with fear as I peer over the cliff. We're all going to die.

But I awaken before we jump to our deaths. Thank God, it was just a dream. I'm not really about to fall to my death like an amateur idiot with no parachute. But that song keeps playing in my head over and over. "Only the Good Die Young..."

My Tom Cruise nightmare is very similar to what I'm planning in real life. People are already looking at me like they don't expect me live much longer. And because I'm in charge of the wilderness search, this will probably end in disaster. My old flying partner Gibbs used to say, "Dave, you could screw up a two-car funeral." And I have to admit, Gibbs was right about me.

I've overlooked a lot of danger here. First, there's the danger of getting involved with Storybook Inn. Larry was right when he said he wouldn't fuck with the "don't fuck with us" gang. Most thinking people know better than to invite this kind of trouble. But I'm inviting everyone to meet me at Storybook Inn to search for a missing hotel guest, in full view of the gangsters who may have murdered him. I'm pretty sure that'll be considered "fucking with us." In fact, I have two days of public events planned with candles and music. I'm surprised my friends said yes to this.

The plan is to park my RV in full view of the criminals at the school bus stop so Randolph can set up a command post for his investigation. Donavan's family can wait in air-conditioned comfort while we search around Storybook Inn. In doing so, I've invited the public to a secret, illegal business that the sheriff captain said is run entirely by criminals. That could be a problem.

Mother Nature is unforgiving. Rattlesnakes don't care that I want to do something nice for Donavan's family. They'll bite me and my neighbors to death.

Gravity doesn't care. It'll drop us to death like my Tom Cruise nightmare. In fact, during this search, gravity could kill somebody with a fall or by hitting them on the head with falling rocks.

The sheriffs failed to solve Donavan's case, so what I have is a bunch of amateurs. Hugh asks Craig if he can land his helicopter and airlift us after a fall or snakebite. Craig says that's not allowed, but in case of life and death, he'll do what needs to be done to get us out of there. Thank God. After two years of murderous corruption, it's nice to have friends who won't leave a man behind.

It's easy to get lost in those woods. I may end up with people missing from the search. It's ironic, but a real possibility.

Even the helicopter isn't immune from danger. Mountain flying is treacherous. Craig and Sam will have to fly above the clouds to get to Sky Forest. That's one of the most dangerous things you can do in a helicopter.

I try to act calm, but I feel like I'm about to jump without a parachute into the abyss below Storybook Inn. If anything bad happens, it'll be my fault.

I get myself into trouble all the time, but now I've outdone myself, and I invited the media. I want the media to get the word out about Donavan's disappearance. If Donavan skipped town, like Storybook workers claim, Donavan might see us on the TV news and get the idea to call his parents. Hopefully he won't see us on the news because we got lost or killed while searching for him.

I've cheated death many times in my work as a helicopter cameraman. But this time is different. This time I can't stop dwelling on what could go wrong. As I sit at my dining room table looking at search invitations and terrain maps, I can feel the stress and fear welling up inside. But then something unexpected happens. I feel a calm, loving presence. And I hear a voice that says, "Don't worry. Just take the steps. It will be OK." I swallow the lump in my throat.

"OK. Let's go."

"TREAT THEM LIKE FAMILY…"

I TOLD DONAVAN'S FAMILY, "YOU ALWAYS have a place to stay in Sky Forest." So they make the long drive from Fresno to my house. As they pull into town, they pass the school bus stop where Donavan disappeared eight months ago.

Dealing with Storybook Inn is no picnic. With tears in his eyes, Shannon says, "We spent all our money searching for Donavan. They had us looking everywhere from Nevada to Long Beach…"

I don't know the Doyles very well. We only just met. But the voice tells me to "Treat them like family," so I clean the nicest room at Cloud Nine and make it ready for them.

As they get settled in the house, I sense they're stressed out and at their wits end over the disappearance of their son. But as we gather to share stories of the sheriffs and the crime ring, I think

I can see them relax just a bit. After arguing with sheriffs and hotel staff for eight months, it must feel good to have a friend who's willing to listen and help them.

I hold nothing back from the Doyle family. I'm even borrowing money to hire the helicopter for Memorial Day. That probably makes no sense to thinking people. But I'm not thinking. I'm just following the voice in my head.

If you think that sounds odd — to set everything aside and go into debt for complete strangers — then I suggest you fasten your seatbelt. You're about to enter a very strange chapter of my life.

LIGHT UP THE DARKNESS

I HAVEN'T BEEN SLEEPING BECAUSE THERE'S so much work to do. There's no end to the phone calls and logistics. I know I'm leaving things undone, because there's only one of me and a lot to do. But the real reason I'm overwhelmed is, I'm not very good at organizing stuff.

The good news is some of my neighbors are coming to help. As the sun sets through the trees in Sky Forest, we meet at the last place Donavan was seen, the school bus stop. I step up to Steve's microphone and welcome our friends and neighbors to the prayer service for Donavan Doyle.

 Help find Donavan Doyle
Liked May 18 %

A Sunset Service on May 25th will take place at 7:30 pm in beautiful Sky Forest at 8:15, a Magic Hour lighting of lights and prayers for the family at the public turn-out at Hwy 18, Kuffel Canyon Intersection, above Willow Woods, where Donavan was seen entering the forest, on the day he disappeared. (battery powered lights will be used, no candles, per fire danger)
The following day (Memorial Day) a search expedition will enter the wilderness to search for signs of Donavan Doyle. Renowned CSI Investigator, Randolph Beasley will comb the area for traces of evidence that may help solve the mystery of what happened to Donavan Doyle, the night he disappeared.
We pray that Donavan will be found alive, but also that his family may receive some love, care, support and answers.

Please join us if you can make the trip to Sky Forest. If not please light a candle at 7:30 and say a prayer for Donavan's safe return.

I would like to personally thank the community of Sky

"Raise your hand if you've ever lost someone," I say.

Hands go up. I guess I'm thinking of the friend I lost last year, David Gibbs, because I'm suddenly overcome with emotion. Gibbs died over a year ago, but for some reason my tears are making it hard for me to speak.

"Raise your hand if you believe in miracles," I say.

Hands go up.

"That's good. I believe in miracles too."

I introduce Donavan's family to my neighbors. Donavan's mom and his aunt step up to the mic and talk about the things that made Donavan special. Donavan's mom says she believes we're going to find him.

Donavan's grandfather is here from Fresno. He's very ill and in a wheelchair. But I'm glad he made it here so he can see that people care about his grandson and are willing to make an effort.

Angela Park wants balloons released with messages for Donavan during the prayer vigil. Unfortunately, Angela's prayer balloons fall flat. We seem to have gotten the wrong kind of balloons because, instead of rising into the sky with handwritten wishes for Donavan, they roll around the floor of my RV. It's a reminder that although well-meaning, we're a bunch of amateurs, and we need a miracle to find Donavan.

Pastor Mack from Sandals Church, Lake Arrowhead steps up to the mic and leads an emotional service with beautiful prayers while Steve plays the guitar.

In the end, we turn off the lights, cameras, and microphones and join hands in a prayer circle. The church next to Storybook Inn refused to help us, but here on the side of the road, we're having church.

This is one of those rare moments when my cranky, opinionated neighbors set everything aside and just love. In spite of the murderous corruption at Storybook Inn, it gives me hope.

"WE KNOW WHERE YOU LIVE..."

"WHO IS WE?"

Back at Cloud Nine after the prayer service, someone I don't know is on the phone. And he's reading one of my emails to me. I don't know this person. I didn't email him. And I didn't give him my phone number.

Sky Falling A Documentary

But he's on the phone and going through my email, line by line, and says he's not happy because I'm trying to expose Storybook Inn.

"Who is helping you!?" he demands. I get his message loud and clear. Anyone who tries to help me or shine a light on the school bus stop can be found and attacked. I try to remain calm as he goes on a tirade, repeating Janice Rutherford's talking points. I guess I should know from this that I'm being watched, and the bad guys will foil any plans I have to expose them.

Intimidation is the biggest reason why my neighbors quit fighting.

I don't like it either. I'm just not willing to give up the fight. I'll quit when the bad guys leave the school bus stop. I'll quit when gangsters stop luring drug addicts into Storybook Inn and threatening to kill the rest of us.

But how did this guy get my emails? It's scary, because I have a 4-year-old son who sleeps next to Storybook Inn. The thug on the phone is making sure I don't feel safe making a public scene at the hotel tomorrow.

But I don't know what else to do. The smart people quit fighting against the gang six months ago. I'm the only one still organizing protests against the crime ring.

The word on the street is thirty-nine people have disappeared just like Donavan Doyle. If that sounds unbelievable, I understand. I never would have thought authorities could conspire with criminals to run a nationwide scam like the one at Storybook Inn.

Click. —the phone goes dead. That was rude and unsettling. I rub sleepy eyes and let out a nervous breath. I know why the guy called. The bad guys want me to stop calling attention to their scam. They're making a hundred million dollars illegally, and they might lose some of their money if people start asking too many questions about school bus safety and missing hotel guests. If I can create a big enough scene, the bad guys might not be able to lure in vulnerable patients like Donavan Doyle.

After the phone call, I sit down with the Doyles at my dining room table to discuss Storybook Inn. Gianny pulls his headset off his head and winks at me. While filming the Doyles, Gianny can hear Wyatt on the sound track playing with his toys.

"Wyatt! Be quiet buddy!" I hush my toddler son.

"I don't like it in here!" Wyatt shouts from behind the door. He protests being locked out of the room, causing all the grown-ups to giggle fondly. I guess hushing him won't work.

Wyatt is only four years old. We spend all our time together, so he's not comfortable being locked out of the room I'm in. But he can't be in the room when we discuss the gangster business. It's hard to fight organized crime as a single dad.

"OK, buddy. Bring your toys and you can play in here, but I need you to be quiet like a church mouse."

The grownups smile as Wyatt obediently moves into the room and tones down his play so we can finish going over the history of Donavan's time at Storybook Inn. I can feel the tension go out of Wyatt's body as he reconnects with me. He feels instantly better when we're in the same room.

I feel weird having Wyatt in the room for this discussion, but I also see in the knowing smiles of Donavan's loved ones that they understand why I'm fighting the people who took their son. I have a bigger problem than solving the mystery of Donavan's disappearance. I'm trying to protect Wyatt from the "don't fuck with us" gang.

The Doyle family gave the "don't fuck with us" gang a king's ransom to get Donavan off drugs. Now the money is gone, and their son is gone too.

Donavan's loved ones grow more concerned with each word I say about the history of crime at Storybook Inn. Sometimes they share knowing looks, as if to say, "That's what we were afraid of..." As they learn the secrets of Storybook Inn, I can see the wheels turning behind their eyes. They're right to be afraid for their son. This is a nightmare.

> **"This is a parent's worst nightmare."**
>
> — Shannon Doyle (Donavan's dad)

How can Donavan disappear without a trace when there are so many people at the hotel? Something's not right with the hotel workers' story.

This missing person case is looking more like a murder mystery. Storybook Inn is looking more like a county-sponsored crime ring. The more we talk, the more I realize that it's unlikely that Donavan just wandered off under his own power like the gang is claiming. There are too many unanswered questions. Renae Pasquale holds up a fistful of emails from the sheriffs promising to investigate Donavan's disappearance, but she says the sheriffs never investigated.

Renae's eyes light up when she recounts how the sheriffs warned Kory Avarell when the Doyles were coming to Sky Forest to search for Donavan. She says, "That's when I knew they (sheriffs) were corrupt." Everyone nods in agreement. No one seems comfortable with the sheriffs lying, refusing to investigate, but then taking the time to warn Storybook thugs that the victim's family is coming to ask about Donavan. The sheriffs are behaving like members of the "don't fuck with us" gang.

There are levels of cooperation and cover-up that would make any parent worry. According to Janice Rutherford, Donavan went missing from a rehab that's not even a rehab. Now what are Donavan's loved ones supposed to do? If the rehab doesn't exist, then what about their son?

Why did the Storybook staff not give details of Donavan's

disappearance? When questioned, they would only say, "He left. We don't know where he went..."

Eyewitnesses say the "don't fuck with us" gang had a fist-fight with Donavan at the school bus stop and chased Donavan into the woods. That was eight months ago. Now the same thugs claim to have no idea what happened to Donavan or where he went. I'm not very clever, but I can clearly see that the Storybook workers are hiding something.

When the Doyles asked, the Storybook staff lied and said, "We don't chase addicts..." But they were seen chasing Donavan.

Nothing I tell Donavan's family of Above it All and Storybook Inn sounds like a legitimate health-care company. The more we talk about it, the more concerned their expressions get. Nothing they say about the sheriffs sounds like legitimate law enforcement.

I've been warned many times not to make too much noise about this because people can be murdered and their homes burned down to keep the money flowing to Storybook Inn. But I made the decision to treat the Doyles like family. And when I look at Donavan's dad, I see myself. When I look at pictures of Donavan, I see Wyatt. That could be me someday, trying to find my son. And I hope someone will open their home to help me, even if it's dangerous to do so.

I wish I could say I'm not worried, but my home can be burned down by the thugs that made Donavan disappear. And I don't know if the sheriffs would investigate.

Sky Falling
A Documentary

Sky Forest is a magic place. I hope that magic keeps us safe in these dark times. I know it's not realistic, but it's all I've got.

It's past Wyatt's bed time, but I still have a lot of work to do. I must excuse myself to go upstairs and put Wyatt to bed. When I say goodnight to Donavan's loved ones, I'm exhausted. It takes every ounce of energy I have to climb the stairs. And that's where I find Wyatt. Sadly, I'm not the only one working past bedtime.

Wyatt has fallen asleep while patiently waiting for me to take him to bed. He's such a good little boy. He is so well-behaved that, while we were talking, he didn't make a sound. I didn't even know that he was a few feet away and hearing every word. As we talked about thugs and felons, my four-year-old son slowly drifted to sleep in the most uncomfortable position I've ever seen.

A toddler shouldn't be growing up around corruption and murder, but I have to fight these things to make him safe again. Unfortunately he got dragged into the fight with me. And now his little body is curled into a pretzel against the hardwood stairs of Cloud Nine.

I hope this book travels far and wide, so the "don't fuck with us" gang can be brought to justice, and we can sleep better in Sky Forest.

"I GOT A WEIRD FEELING."

"David, I prayed about Donavan and I saw a ditch with a stream running through it," Angela Park says. "I think you'll find something there. And I think something bad happened in a park nearby."

Angela says she prayed for us and got a feeling there is a park nearby where something bad happened to Donavan. Donavan's mom said she has a weird feeling about Switzer Park, where one of the hotel workers hunted Donavan. Hugh Campbell also said he has a strong feeling about Switzer Park. As each person comes to me separately to talk about weird feelings, this doesn't feel like a coincidence.

Because so many people have strong feelings about Switzer Park, I put it on the list of places to be searched by volunteers and helicopter.

JANICE'S ROLE

As I'm preparing for the wilderness search, I look down at my phone. A New York number is calling.

"Hey, Charles," I say.

"Hey, Dave. I got bad news. Are you sitting down?"

"Uh oh."

"I just got a note from Janice Rutherford's office. They're cancelling the interview."

What the hell is this? Janice Rutherford suddenly cancels her interview for our documentary?

Charles fills me in on the details. Rutherford's staff agreed to the interview, but now says, "Due to impending litigation, Supervisor Rutherford won't be commenting on Storybook Inn."

Impending litigation? That wasn't a problem before we announced the search for Donavan Doyle. Janice sounded so confident about her position on Storybook Inn. What changed?

Litigation never stopped Janice Rutherford at Storybook Inn before. In fact, she threatened us with lawsuits over Storybook Inn. Now she's afraid of lawsuits? That doesn't add up.

The Doyles aren't suing anyone. Does Janice Rutherford know something we don't know about Donavan Doyle? This situation is getting stranger by the minute…

I made arrangements for Charles to travel from New York and for Gianny to bring his world-class cinematography to give her respectful treatment in our film. But now she's backing out. You know what this reminds me of? Kory's no-show meeting at the school bus stop. Kory and Janice both agreed to speak about Storybook Inn, but when push came to shove, neither of them showed up. That makes me feel like everything they said about the place is a lie. But…that's just my opinion.

Now that Janice Rutherford has refused to answer questions on the eve of a search for one of the victims, I'll let you draw your own conclusions about her role in Storybook Inn.

I hope you're reading this, Janice. I have a growing list of questions for you about Storybook Inn, Santa's Village, and Donavan Doyle.

MEMORIAL DAY

THE EARLY-MORNING SUN FILTERS THROUGH the trees into the windows at Cloud Nine.

It's one hour before the search. Gianny sets up his Red camera and a shotgun microphone. Shannon Doyle looks tired. He draws a tense breath and talks about the shady way rehab people and sheriffs hid information and refused to help his missing son. It's maddening. Shannon and his wife have been stonewalled for eight excruciating months. Suddenly he bursts into tears. He begins talking as if Donavan is no longer alive. It's very upsetting.

I don't have to look around the room to know that everyone is affected by this. I'm trying to be cool and professional, but I have to wipe tears from my own eyes.

Maybe I take Shannon's suffering personally because I'm a dad. I don't know.

I'm doing my best to keep things organized, but I sense that this situation is erupting out of control. Just like the volcano, there are violent forces bubbling toward us.

Sky Falling
A Documentary

All I can do is "treat them like family." So I give Shannon the best of everything I have, including a brand-new pair of gloves and an aluminum pole hook. The pole hook is important because Hugh said all search volunteers must have a stick for probing bushes for rattlesnakes. I'm using a piece of wood I found in the forest. But Shannon will have a shiny, extendable pole with a prod for pushing and a hook for pulling in the weeds where rattlesnakes and clues to Donavan's fate may be hiding.

AIRBORNE

Hugh Campbell scratches his chin and talks about the mystery of sheriffs at Storybook Inn. "Hey, Dave. Remember when the sheriffs showed up to protect Storybook Inn from questions?"

Hugh is right. When the neighbors came to ask questions about the illegal business, the sheriffs came to stop it. When Storybook owners cancelled their meeting, sheriffs ordered us off the property. The fact that school children are in danger made no difference to them. The sheriffs acted like private security guards for the illegal business.

The sheriffs have million-dollar helicopters and flight crews paid for by our taxes. They could help us find Donavan. But so far I've heard nothing from them.

The sheriffs showed up in force to protect the owners of Storybook Inn when the community gathered to ask questions last year. But I haven't seen them lift a finger to find someone who went missing from the very same business.

There are signs of foul play in Donavan's disappearance. In fact, there are neon billboards of foul play. I hope the sheriffs bring their manpower and equipment to help us investigate today.

A hundred miles away, our helicopter pilot Craig Dyer opens an inspection hatch on a tiny helicopter.

Craig has 28 years' experience flying military and civilian helicopters, but today's mission is different. Today he's volunteering to help search for Cyndi and Shannon's son. And Craig is teaching his own son how to fly.

"OK. Check the blades for cracks. We don't want to find out in the air that there's a problem with the tail rotor," he says. Sam watches his dad check every part of the aircraft before takeoff.

Sky Falling
A Documentary

Craig looks up thoughtfully at clouds swirling above the control tower. "That's not good." Craig's company lost a helicopter flight crew in clouds like these. The pilot got lost in the clouds and crashed in a mountain near Sky Forest. Craig knows that mountain flying is dangerous, but mountains and clouds are worse...

On Cloud Nine I take in a spectacular sight. There's a white carpet of clouds below my feet.

These are the same clouds swirling above Craig's head, but they look different up here. Down in the city, the clouds look grey and gloomy, but up here the view is magical. The tops of the puffy-white clouds are lit by the sun. It's like standing in a surrealist painting.

Sky Forest clouds are some of my favorite things on Earth, but I don't know if Craig and Sam will be able to fly through them to get up here and help us with the search.

Down, below the clouds, Craig and Sam go through the engine start checklist.

"Fuel?"

"Check!"

"Hydraulics?"

"Check!"

Sam is a college student, but today he's learning things they don't teach you in school from a man who's flown around the world.

Craig flew elite Army missions in Iraq ("the Sand Box.") Now he teaches his son to fly a different kind of air support in a machine with a tiny fraction of the power of Craig's Army Blackhawk.

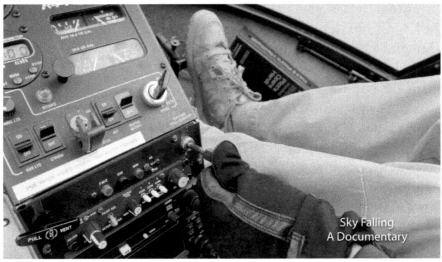

"OK. Now when you start, you have to watch the RPM, so you don't overspeed."

Craig shows Sam how to adjust fuel, air, and ignition, and he jolts the six Lycoming cylinders to life. The rickety machine shakes and rattles like a piece of farm equipment—very crude when compared to the jet-powered squadron operated by the San Bernardino sheriffs. The sheriffs fly three-ton helicopters with high-tech, turbine-powered blades. But Craig flips a switch that squeezes a rubber belt against a metal drum to spin the rotors. Inside the R44 engine compartment, it looks like a lawnmower and a bicycle made a baby as thin metal tubes and rubber belts spin a hamster-wheel contraption to spin up the blades.

The ultra-light R44 is a far cry from the advanced heavy helicopter that crashed in our mountains killing three of Craig's colleagues December 10, 2006.

God bless Craig and Sam. They're flying into the same perilous environment that took the lives of a flight crew who had the biggest, most powerful aircraft that money can buy.

"Tower, Helicopter Four November Lima...ready to lift..." Craig asks permission to take off and begin the journey to Sky Forest.

The tower answers back, "Helicopter Four November Lima, altimeter two-niner, niner-eight...cleared for takeoff."

Craig turns to Sam. "OK. Get ready. Follow me on the controls."

Sam lightly grips the collective lever and cyclic stick. He touches his feet to the pedals so he can feel how his dad controls the helicopter.

The air is charged with adrenaline and anticipation as the little Lycoming engine creates a mad-dash drumbeat with six boxer cylinders pounding at two thousand seven hundred RPM. Craig carefully watches the horizon and pulls the collective with his left hand, while he pushes with his right against the cyclic stick, and his feet to the pedals in anticipation of what comes next.

Sam can feel his dad moving all four levers simultaneously with the experience of a thousand hours in military and civilian helicopters. They sink in their seats as Craig pulls the collective and they become light on the skids.

This is a critical moment. Before Craig adds power to the blades, he feels carefully for uncommanded movement. When he's satisfied that every piece of the thunderous little machine is fully under his control, he gently pulls up on the collective and pushes the cyclic stick to lift.them into the air. Craig pushes his foot against a pedal and they spin toward Sky Forest.

Craig switches a radio to the frequency on my handheld walkie and broadcasts, "Raptor helicopter is airborne."

PANCAKE MOVIE

Six thousand feet above Craig and Sam, in Sky Forest, Gianny moves a hundred pounds of movie equipment into my kitchen. With the intense focus of a master craftsman, he builds his Red camera and a telescopic lens.

"What are we doing, Dad?" Wyatt asks.

"We're making the pancake movie, buddy," I say. Gianny grins at how I relate our documentary about organized crime to my four-year-old.

It makes me nervous to have Storybook Inn victims in the house with Wyatt, but I'm "treating them like family." So here we are about to eat breakfast.

Wyatt is supposed to be with his mom today, but she left him at my house. I've been very careful not to expose Wyatt to gang activity at the Storybook Inn, but he's here as we take on the "don't fuck with us" gang. Wyatt will be here when we march right up to the gang headquarters and ask the gang what they did with one of the victims. Wyatt will be here when our helicopter crew hovers over their building and snoops in their backyard. I shouldn't bring a four-year-old into a battle against organized crime. But the sheriffs aren't helping. And someone has to help the Doyle family.

Like an erupting volcano, I don't control the violent forces at Storybook Inn. But I control what goes into my pancakes. For a few minutes I feel relieved as I cook for my family. Gianny turns a wheel to focus on blueberries and honey mixed with whole-wheat flower. I could use store-bought pancake mix, but homemade pancakes are better for Wyatt.

As strangers gather around the big table at Cloud Nine, Wyatt pulls blueberries out of his pancakes and eats them one at a time.

After breakfast we'll head to the hotel. We're going to investigate the "don't fuck with us" gang in full view of the gang.

Sky Falling
A Documentary

As Wyatt devours his pancakes like a hungry bear, I worry about what may come and how it will impact him. Wyatt's grandpa will have to watch him while I climb into the forest behind Storybook Inn and stir up the hornet's nest...

SEARCHING ABOVE THE CLOUDS

"Wˈʀᴇ ɴᴏᴛ ɢᴏɴɴᴀ ᴍᴀᴋᴇ ɪᴛ, Dave." Craig is on the phone and says he can't fly through the clouds with the helicopter. They're stuck at an airport down in the city. I take a deep breath. This is bad news.

The clouds are beautiful, but also deadly. When Craig and Sam climb toward Sky Forest, the clouds block their cockpit view. A condition known as whiteout blinds them. This is the same condition that caused the tragic helicopter crash that killed everyone on board in 2006. Fortunately Craig aborted his mission before anything bad happened to him and Sam, but now we'll have to search around Storybook Inn without air support.

But we already know there are areas that can ONLY be searched by helicopter. Maybe the sheriffs will send a helicopter from their Mojave base. They're in the clear and could be here in minutes. But so far I haven't heard from them.

Anyway, it's time to go. I load up the RV on schedule and drive to the school bus stop to build our command post for the Memorial Day Search. I hope to see news vans and sheriffs' cruisers there to help us.

If Donavan wandered off and is living on the street, like the Storybook staff says, he might see us on the TV news and call his loved ones.

Hopefully the sheriffs will help keep an eye on the volunteer search and swing into action if we find something in the woods behind Storybook Inn. I still have hope for the sheriffs. I told Donavan's family, "Mark my words, the sheriffs will be standing right beside you on Memorial Day." I admire our sheriff deputies and hope they'll come through in spite of the crooked politicians that fund their paychecks.

But when I turn my little RV into the school bus stop, I find nothing but an empty lot. The sheriffs aren't here—just me and my friends. The media isn't here either, just Gianny and his camera.

I park the RV, fold out the awning for shade, crank up the generator, and switch on the air conditioner for Donavan's family. One by one, my friends and neighbors gather and prepare to search the mountain. I try not to think about the danger around us. I take a nervous breath. Searching around gang headquarters will bring us into harm's way. It's like hovering a helicopter over a volcano.

As usual, I don't have a very good plan. But my friend Hugh Campbell has planned everything down to the inch. He neatly folds custom-made terrain maps and starts programming walkies so we can keep in touch as we spread out in the forest behind Storybook Inn. Thank God Hugh is here.

Hugh's background as an aerospace engineer is showing today. He's engineered this search to perfection. He points to his topographic map and says, "I know Dave would like to go all the way, but I'm being more conservative."

Hugh is talking about my crazy plan to climb from Storybook Inn all the way down to the city. But my plan makes no sense when we look at Hugh's map. So Hugh points to specific areas that each of our volunteer search teams can realistically search.

Donavan's loved ones have become bitter toward first responders after almost a year of lies, cover-ups and stonewalling. I can't blame them. I've seen more effort to find a stray cat than they put into finding Donavan Doyle. And now Janice Rutherford has cancelled her interview. Public officials aren't helping the victims of Storybook Inn, and they're refusing to talk about it. If they did this to my son, I don't know what I would do.

The only time I see the Fire Dept. or sheriffs at Storybook Inn is when the "don't fuck with us" gang wants something. When the hundred-million-dollar scammers want free ambulance service or private security, the Fire Dept. and sheriffs are there in minutes.

I've seen three sheriffs' cruisers at a time at Storybook Inn. That doesn't feel right. Sheriffs are supposed to protect families in our community, not work as private security for an illegal business.

I suppose everyone knows this is illegal and a misappropriation of public safety resources because, to this day, the sheriffs' staff denies ever going to Storybook Inn. Hugh has been warning me to pay attention to how sheriffs handle Storybook Inn.

But Hugh doesn't share my cockamamie goal of charging

straight into the wilderness behind the hotel. As leader of our ground search expedition, Hugh is preparing a safe plan for our volunteer teams.

"It's possible to make it all the way down to the city," says Hugh. He points to the junction of three canyons below Storybook Inn. "We're going to know when we make it here if we want to continue."

Hugh and I decided that all volunteers can search the school bus stop and Storybook Inn. But only those with climbing experience can venture below Sky Forest Springs, where the terrain becomes near-vertical. It's the exact spot of my Tom Cruise falling nightmare. Only expert climbers are allowed to venture down there.

Each team will be headed by an experienced outdoorsman who can keep us safe and accounted for. The one thing I'm trying to do is keep track of our volunteers. So I asked team leaders to be responsible for roll call at the end of the day. I don't want anyone left behind that stupid hotel after our search.

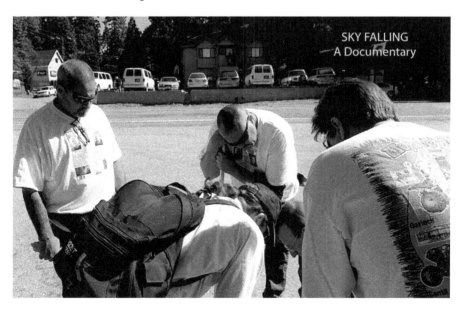

There's an eerie tension in the air as we look at maps of Storybook Inn in full view of the Storybook thugs. Instead of the spirit of recovery, I sense fear and anger from the "don't fuck with us" gang. I think they know their illegal business will be harder to run after today.

Gianny turns a metal ring and focuses his Red lens on the bus stop building, where gangsters shoot daggers with their eyes. The last time our cameras were here, Kory Avarell and a gang of bad guys came out to prevent the camera crew from filming the school bus. But today I brought cameras right to their front door, and I invited the community to search for one of the victims of their scam. I have to accept full blame for this one. There are few who would elevate "fucking with us" to a public art form.

When creating massive problems for myself, I usually start with a small idea. I almost didn't survive my trip to Peru. At first I just wanted to take care of my camera, but I ended up a wanted man. In Sky Forest, I just want to help a family in need. So we're at "don't fuck with us" headquarters with an RV, cameras, and a crowd of amateur investigators.

As Donavan's mom and dad walk over from my house, a volunteer asks Shannon about the day his son disappeared. Shannon rubs his tired brow and recounts the conflicting stories he got from hotel workers and the sheriffs. None of it adds up. I don't want to think of it, but I have a bad feeling about what we may find today.

Hugh divides our volunteers into two teams. He'll lead his team to Switzer Park, and I'll lead my team to Sky Forest Springs. This will make the best use of the sparse information that's leaked out of Storybook Inn and the Sheriff's Department.

I twist a knob on a walkie and press the PTT button. Hugh's radio beeps in response. Good, we're linked. But across the street, Storybook thugs are staring at me like I'm a piece of meat. The tension is as thick as the clouds below my house. I'm afraid this may be too much "fucking with us." The volcano of corruption might blow up in my face at any second.

I hope the Doyle family will leave town having found nothing but love and kindness in Sky Forest. But the way those thugs look at me, I feel like we may not survive this. My throat tightens as the bad guys stare. This is why some neighbors ran from the fight. The "don't fuck with us" gang rules our neighborhood with fear. I think that's why no one really knows what goes on at Storybook Inn. People are afraid ask questions or speak about what they know. I'm not immune from the danger. When the Doyle family leaves town tonight, Wyatt and I will have to sleep next door to the thugs who are mean-mugging me right now.

CSI

RANDOLPH BEASLEY BECKONS OUR SEARCHERS to gather around him. He explains what human remains will look like after eight months behind Storybook Inn. He says, "Bones will look like sticks, so be careful where you put your feet." He hands out strips of crime scene tape. "If you find something, don't pick it up. Don't touch anything. Take a picture. Mark the location with my crime scene tape and call me on the radio. I'll come down and take control of the scene. Do not touch anything!"

"I'M SURROUNDED BY ASSASSINS..."

As villains hunt me with their eyes from across the school bus stop, it reminds me of a hardened killer I faced years ago. I was minding my own business, bolting my camera to a helicopter, when he came at a high rate of speed in his notorious Jeep, homing in on me like a heat-seeking missile. Dramatically screeching to a halt, he exits the vehicle like a death-eating commando and creeps around the car—as if I can't see him there. My own killer instincts take hold and I find myself crouching behind the helicopter preparing a violent ambush for my would-be-assassin.

Through the tinted windows of his jeep, I see the face of death stalking me. I match his evil stance. We are both ready to strike...

Suddenly, his mouth contorts in an unexpected fashion. Is he smiling at me? What the hell? I feel an involuntary smile creep across my own expression. I can't help smiling at my friend of eighteen years who stalks me every day of my working life. Why can't David Gibbs just show up to work like a normal person? I try not to smile, but I can't help it. His onslaught makes me laugh out loud. I feel sorry for him, so I give him a handshake. And because I love him, I take a minute to straighten him out. "Number two! You're early. Get out of here."

My pilot David Gibbs is always here hours before he needs to do pilot stuff. The camera's not built yet and there's nothing for him to fly. So I order him away.

Shaking his head, he says. "No. I'm not going to leave you

unsupervised. Get me my wrenches." And just like that, he steals a fistful of my tools and starts working on my camera. Gibbs parades across the hangar like he owns the place, digs into my camera cases, finds a part, and starts bolting it onto the aircraft with my wrenches. And then he demands, "Peanut-head! Get me my music!"

Who does he think he is, showing up early and barking orders? Now he's waving his hands in the air like an orchestral composer, eyes closed, ready to hear the sound he's conducting. This is unbelievable.

So I put on some music, but not the song he wants to hear. I'm not going to give him the satisfaction.

"Peanut-head! That's the wrong music. Get me my *Pirates of the Caribbean!*"

Gibbs' favorite song is from the movie sound track, "We are Pirates." But I'm not going to play it for him. That would be giving in to his terrorist demands. So I put on Britney Spears instead.

He stares at me as if he's a Tyrannosaur about to eat an insect. Apparently he noticed I'm not playing "We are Pirates," but some other song that I picked. He doesn't blink…just stares at me…waiting…

He breaks off his murderous look and dramatically complains to no one in particular, "I'm surrounded by assassins!"

In eighteen years, I never had one normal conversation with Gibbs. I miss our daily arguments. And I wish he were here to steal my tools and give me a hand. But Gibbs died in a helicopter crash last year. Now I'm fighting an organized crime ring at a school bus stop…and I'm surrounded by assassins…

BARELY READY

Our volunteer chaplain is here, but there's a problem. I can't take my eyes off Robert's bare feet. He claims he's hiked entire mountains without boots. I don't know what to say, except, "Did you bring shoes?"

Robert laughs and begins lacing up a pair of hiking boots. He also hands out his own walkies to search volunteers.

I scan the crowd but can't find my toddler son anywhere. To my horror, I find Wyatt wandering away from the search party. "Grandpa! I need you to keep an eye on Wyatt! That's the most important thing right now!"

Grandpa is arguing about how the cars should be parked so

we don't get parking tickets from crooked cops who might punish this uprising. That's great; I don't want a parking ticket. But grandpa is supposed to be keeping an eye on Wyatt while I search for Donavan.

My toddler can't go with me into the forest, and I need to know that my little one is safe while I'm looking for Donavan. One missing kid in Sky Forest is enough. Renae Pasquale sees the problem, walks over to Wyatt, and says over her shoulder, "Don't worry. I got this." Thank God. She wrangles Wyatt and Grandpa so I can get back to work on the search.

We set Donavan's mom up in the air-conditioned RV as Donavan's dad and I prepare to enter the forest. My team will descend the slippery slope from my nightmare below Sky Forest Springs. The other team will search Switzer Park.

Donavan's mom, Hugh, and Angela said they have a strong feeling about Switzer Park. Donavan's dad says he had a strange experience at Switzer Park when they drove into town yesterday. Uh oh.

I have a strong feeling too, but it's not about Switzer Park. My feeling is hard to explain. There's a dialogue going between me and someone I can't see. Remember when I heard a voice in the Calgary airport? It's happening again right now. When I worry about snake bites, falling rocks, and volunteers getting lost in the forest, I hear a voice that says, "Don't worry. Just take the steps. It'll be OK."

"OK," I say. "Show us something."

That's strange. I have no right to ask someone I can't see to show me something. I don't even know who I'm talking to.

Robert shows us something. He picks up a giant earth worm that was slithering across the path in front of us. The fat worm leaves a trail of slime as it wiggles across his hand.

"That's disgusting," I say.

The big, slimy worm feels like a bad omen and a terrible way to start a search for a missing person. A feeling of darkness comes over me.

Robert puts the worm back onto the earth. The earthworm is a symbol of death and decay—not what I'm hoping to find in front of Donavan's parents.

Suddenly, a thought hits me like a bucket of cold water. "Robert! We forgot to say a prayer!" I yell.

The barefoot chaplain doesn't hesitate. He grabs one of his walkies and swings into action. "Listen up all groups as we say a prayer." Throughout the forest, my friends and neighbors stop in mid-stride. They all lean in to hear Robert's voice over crackling walkies.

"Lord, you know all things. Bring us to Donavan, Lord. Show us where he is." My walkie chirps as each volunteer says "Amen."

After the walkie-chorus of "Amens," we continue our march into the wilderness.

Robert's stepson David Wemyss carefully looks through a thorn bush for any sign of Donavan Doyle. The bush is almost impenetrable, with two-inch thorns that look like wooden hypodermic needles. This is the same kind of bush that took apart my drone and stuck me on the scout mission. But Wemyss navigates the thorny bush with ease. At age 16, he's the youngest and, surprisingly, the most capable volunteer we have. His pouches and pockets jiggle with wilderness survival gear, a snakebite kit, a knife, and a walkie-talkie.

A cross-country runner, Wemyss bounds up the mountain like a spring-loaded spider monkey. He has no problem with the heat and altitude that's already slowing the rest of us down.

RIGHT PLACE, WRONG TIME

CRIME SCENE INVESTIGATOR RANDOLPH BEASLEY looks down at his wristwatch. It's 10:00 a.m. His walkie crackles with radio calls as amateur teams descend the treacherous mountain below the RV command post.

Sky Falling
A Documentary

Randolph prepared our volunteers as best he could, but not everyone attended his CSI class. One of my neighbors walks over to the school bus stop and makes a terrible discovery.

Storybook Inn workers entered the forest two hours ahead of us. They didn't tell anyone where they were going. They skipped Randolph's 9:00 a.m. CSI class. Hotel workers walked into the forest, on their own, at 8:00 a.m.

"Why would they do that?!" Randolph asks. A pained

expression comes over him. But I know why they did it — because I'm an idiot. I allowed the search invitations to be mailed with the wrong start time.

We mailed invitations to the 10:00 a.m. wilderness search to every PO Box in Sky Forest. But the invitations said 8:00 a.m. — two hours before the search was scheduled to start! It's a stupid, amateur mistake. Now we know this is a David Alan Arnold production.

Our volunteers showed up on time for the 9:00 a.m. CSI class because we'd discussed the schedule in person. But some volunteers didn't talk to us, because they work for a secret, illegal business the county regulators say doesn't exist.

Storybook people only knew the time written on the mailed invitation. So they came at 8:00 a.m., just like is said on the paper I sent them, and when they didn't see anyone else, they began searching on their own without CSI training. They entered the forest without supervision or a safety plan.

By the time Randolph began his CSI class, Storybook workers were already deep in the forest and potentially doing everything he warned us against — stepping on things, removing things, disturbing physical evidence and clues to Donavan's disappearance.

I absolutely want Storybook people to help with the search effort. But I didn't foresee that my mistake would cause them to operate without any connection to our team leaders and without Randolph's expert supervision.

I don't even have a headcount on their team. If they get lost or hurt, I don't know who they are or where they went. They don't have walkies to keep in touch with us. We can't share information or offer assistance in case of emergency.

And Randolph is right to be concerned. The same people who fought with Donavan and chased him the day he disappeared might be the ones who are now poking around in the woods behind the hotel. If they find something, they might contaminate it or destroy it. We could lose irreplaceable evidence of what happened to Donavan.

FIRST FIND

MEANWHILE, I'M STARING INTO THE abyss of my Tom Cruise nightmare. The drop-off looks nasty. I hope no one falls down there.

Unlike Wemyss, I don't have cool holsters or pouches. What I have is a school backpack jury-rigged with Nathan's quadcopter toy. It's ridiculous, but it's the only way I can carry the little flying machine to keep my hands free for climbing.

Sky Falling
A Documentary

It's hard to see things in the thick forest, and the drone is already helping me. Donavan's family can watch the video later and search the forest with their own eyes from the comfort of my RV or their home in Fresno.

"C'mon little quadcopter. Do a good job," I say as Gianny holds it above his head. *And please don't crash.* Gianny cringes as if he's about to get attacked by hornets. I haven't even lifted off yet, but he's already bracing for the impact of the spinning blades.

I must admit, I'm a little bit surprised when it doesn't crash. Thank God it lifts perfectly out of Gianny's trembling hands and zips above us. In seconds, it effortlessly floats to places that would take an hour to reach on foot. And so far I haven't hit any trees or run out of battery like I usually do.

This footage is priceless. We can scour the drone footage to see if Donavan left any trace of evidence when he disappeared.

I hope Donavan is still alive, wandering the world, like the Storybook gang claims. But just as we're starting our search, my walkie comes to life with multiple calls, including ground search leader Hugh Campbell, who says, "Wemyss found a piece of evidence." Uh oh…

I land the quadcopter, climb above Sky Forest Springs, and find sixteen-year-old Wemyss tying a piece of Randolph's crime scene tape next to a torn piece of human clothing.

CRIME SCENE

Randolph Beasley has had some amazing moments in his CSI career. He won an FBI award by finding a fingerprint inside the pages of a book. Who would imagine a killer would stop to read a book while committing a brutal murder? I don't know. But Randolph was recognized by the FBI for the extraordinary attention to detail that solved a cold case.

Randolph was being charitable when he said we have zero chance of finding anything today. But God bless him, he came to support Donavan's family. While the Sheriff's Department has helicopters, all-terrain vehicles, and highly-trained teams of mountain climbers, we have sticks and a little plastic quadcopter sticking out of my backpack.

Randolph set aside his campaign for Janice Rutherford's office to come out and show some love and kindness to Donavan's mom and dad. After teaching our amateur search team CSI basics, he stands beside Donavan's mom at the RV command post when his walkie starts crackling with reports of "evidence" in the forest.

Randolph strains to hear the garbled chatter from amateur walkies. He presses the Push-to-Talk button. "Say that again, Dave?"

"We're photographing the clothing, and we're going to launch the quadcopter," I say through the static.

A look of befuddlement comes across Randolph's face. I don't blame him. He said he didn't expect us to find anything. But now Randolph glances nervously at Donavan's mom before pressing the button on his walkie again. "Copy that. I'll come down…"

Randolph has investigated thousands of crime scenes, but he's never done it in front of the victim's mother.

Down in the forest, I photograph the clothing just like Randolph taught us, without touching anything. Wemyss carefully marks the spot with Randolph's crime scene tape. Not bad for amateurs. I'm feeling good about our work. But Wemyss is right. This looks like Donavan's clothing. I don't like it.

It's a relief to see Randolph climbing down with nineteen years of CSI experience. He confidently takes charge of the scene and surveys the site.

"I have a bad feeling about this," I say as I point to the chewed-up clothing.

"Really?" A look of grave concern comes over Randolph's face.

Our walkies chirp to life again, and Robert Larivee says, "Ask CSI if he has a fingerprint kit."

Randolph pauses with a look of astonishment. "No. Tell him no, I don't have a fingerprint kit."

Randolph is kind to spend his holiday helping us, but he didn't expect a bunch of amateurs to find anything worthy of fingerprint analysis. His look of shock evaporates as he gets down to business. Snapping latex gloves on his fingers, he looks around methodically and asks if we touched the clothing.

"We didn't touch anything," I proclaim. Randolph taught us well.

As he leans in for a look, his voice becomes energized and tense. "You've already photographed this, correct?"

"Yes, I took pictures," I say as my friendly neighbor transforms into a hyper-alert crime scene investigator.

Sky Falling
A Documentary

Randolph's body language shifts into high gear. His movements become mechanical and precise as he moves the tall grass and carefully grasps the torn fabric with a gloved hand, slowly lifting it into the daylight. For a moment, Randolph stares intensely at the ragged clothing. His mood intensifies. "This is very suspicious," he says. "You see this staining, it has a reddish-brown tint...there's more than enough here to get a DNA profile..."

"Oh my God!" Gianny gasps while focusing his camera on the bloodstains.

I never met Donavan, but my heart sinks as I look around the forest where Donavan might have been murdered. I was hoping Donavan simply ran away, like the Storybook workers said. But our crime scene investigator is doing bloodstain analysis on what appears to be clothing Donavan wore when he disappeared from the hotel.

Sky Falling
A Documentary

Randolph explains, "This is huge. The fabric matches the clothing Donavan was last seen wearing." He gazes thoughtfully over the mountain forest. "I would think that if his body were here, that we could find it..."

God bless Randolph Beasley. He's the only representative of law enforcement on the scene. Thank God he volunteered to help us. If Randolph is right, we're going to need him more than we knew.

I'm strangely unprepared for what's happening right now. Remember when Hugh said I don't know what I'm doing? Well, he's never been more right. My head is spinning. I'm trying to get a grip on what it all means. But Randolph is already scouring the area for more evidence, the way he's done in a hundred murder cases.

Suddenly my phone comes to life. The searchers from Storybook Inn have returned to my RV with pictures of Donavan's shoes and sweatshirt. This is my worst nightmare. If Donavan's blood-soaked clothing and shoes are in the woods, it can mean only one thing. Donavan Doyle didn't survive his stay at Storybook Inn. Now I only have one thought in mind. Someone has to comfort Donavan's parents.

I sadly make the call on walkie. "Does anyone have eyes on

Shannon Doyle?" A member of Shannon's search team answers back from Switzer Park, where they're on their hands and knees crawling through thorn bushes.

"Please bring Shannon to the command post," I say. My throat tightens. I'm not qualified to do this. But I start climbing to the RV to do what the sheriffs should have done eight months ago. I'll comfort Donavan's mom and dad and tell them the truth about what happened to Donavan at Storybook Inn.

The blood-stained clothing tells me that Donavan didn't get away from Storybook Inn like we hoped. And if Randolph is right, we're about to find a dead body behind the hotel.

Sky Falling
A Documentary

"YOUR SON IS LOVED."

My airband radio crackles to life with the sound of Craig Dyer's voice, "Raptor Helicopter is inbound to Sky Forest." Craig and his son Sam are climbing through the clouds. They're coming to help us.

It's a comfort to hear Craig's familiar voice on the radio and to have another pro on scene.

"Roger," I say. "Stand by for CSI. We have bad news."

I give Randolph the airband radio so he can brief Craig on what we found above Sky Forest Springs.

The thunder of Craig's helicopter blades echo through my neighborhood as I approach the RV and find Donavan's dad. I'm not trained in grief counseling, but I know what I must do. When our eyes meet, and without saying a word, we both burst into tears. I didn't expect to have this kind of emotion. But we both sob uncontrollably. I hug him and offer the only thing I have to give. "When you hear those helicopter blades, know that your son is loved…"

It's not much, but at least I can show Shannon that, Donavan is loved because my friends cared enough to fly a helicopter and search for him.

Craig and Sam Dyer circle Storybook Inn with the helicopter.

Sky Falling
A Documentary

Craig looks down at the creepy building where Janice Rutherford says hotel rooms are available.

At this exact moment, Storybook Inn is featured in drug rehab ads across the country. Storybook Inn is also featured on a major rehab website. Their website claims that the hotel has been on A&E's *Intervention*. On *Intervention*, drug addicts receive drug rehab and detox, even though Janice Rutherford says it's a hotel. Meanwhile, county regulators tell me there's no hotel license. God help rehab patients and hotel guests who walk in unaware of the "don't fuck with us" gang.

Craig looks down at where Donavan's shoes and clothes are. It will take us an hour to climb there on foot and search the surrounding forest.

There are no sheriffs onsite to take command of the scene or offer comfort to Donavan's family—just me and my friends. Storybook Inn has an army of tattooed thugs who masquerade as drug rehab counselors, but not one of them are counseling the Doyle family. I've seen more care and concern for missing cats. In the vast hundred-million-dollar business at Storybook Inn, not one company representative has offered to assist the Doyle family, even though I'm told the entire company is aware of the search today.

There's something strange about a hotel with missing guests. Janice Rutherford said it's a hospitality business, but I don't sense hospitality at Storybook Inn. I sense dark secrets fiercely guarded by a crew of vicious thugs that look out of place under a "hotel" sign with little purple hearts.

The sheriff captain said, "They're all criminals." And now the body of a hotel guest is hidden behind their building. I suspect there are other bodies in these woods.

If you know someone who needs help with addiction, please don't send them here. If something bad happens in that hotel, I don't believe authorities will assist you. And I wouldn't trust my safety to hotel workers who brag on Facebook about going to prison.

THE VALLEY OF THE SHADOW OF DEATH

A FTER COMFORTING SHANNON, I STEP into my RV and find a Storybook worker who says his name is Sean.

Sean peers at me through thick glasses. His camouflaged pants and decorative shirt make him look like a nightclub commando. There's a picture of a sub-machine gun on his ball cap and a big, serrated knife strapped to his hip. This is a hotel worker?

Sky Falling
A Documentary

I offer him a cold drink. He politely takes a sip of Gatorade and begins popping pills a coworker brought over from the school bus stop. But, as he lifts them to his lips, one of our search volunteers, Renae, shouts at him, "I wouldn't take those together!"

"What?" He stares at the pills, dumbfounded. Renae says she recognizes the meds by their color and pill shape and cautions him not to combine the powerful drugs.

"Oh," he says gratefully. "I don't even know what these are."

"What?!" How can a healthcare worker gobble narcotics this way? How can he take a cocktail of powerful drugs and be responsible for the care and safety of patients?

It makes me wonder how many people have disappeared from Storybook Inn. And where exactly is Donavan?

"Ex-drug-users run the place and they don't look sober."

— Donavan's Aunt Cheri

I ask how long it took Sean to locate Donavan's shoes and sweatshirt in the woods behind the hotel. Sean says it took forty minutes to climb down to Donavan's things and an hour and twenty to climb back up.

Houston, we have problem. Sean has only been gone from Storybook Inn for two hours.

Sean began his search early because I put the wrong time on the invitation. Had he known the correct time, he could have just blended in with our search team. We might not have realized that he knew where Donavan's shoes and clothes were. But instead of mixing in with our teams at 10:00 a.m., Sean went at 8:00 a.m. and did something very interesting. He ignored all the places Storybook workers told us to search for Donavan. Sean didn't wander the wilderness as we are doing.

Sean and his crew marched downstairs and went directly behind their building. He returned with pictures of Donavan's shoes and clothing. In eight months, not a single Storybook worker told us to look back there.

And Sean made it down there in just the time it takes to climb down and back. That leaves no time for searching through miles of forest between Storybook Inn and the city. Sean didn't have time to search a thousand places Donavan could have gone if he'd walked under his own power.

If Donavan wandered freely, Sean would need time for a needle-in-the-haystack search of the entire forest. Donavan's dad and I are covered in filth from climbing through thorn bushes that cover a hundred square miles of National Forest.

But Sean didn't search the thorn bushes. When Sean entered that hundred square miles of San Bernardino National Forest, he went directly to the spot where Donavan's clothes and shoes were waiting.

Now Sean says he knows details of Donavan's disappearance. Really? Why is he telling us this now, and not in the previous eight months when Donavan's loved ones were begging for someone to tell them what happened to their son?

Randolph catches Sean mixing up dates. "You're talking about today?" Randolph looks up from his iPhone note-taking. Our CSI investigator stares inquisitively at the knife-wielding, pill-popping rehab worker.

Sean pauses and stares blankly at Randolph. The story doesn't add up. Right now, Sean is confusing 2013 and 2014. He points to the school bus stop and says he chased Donavan from there. In the same breath, he also says he found Donavan's shoes and sweatshirt. When Sean speaks, he doesn't separate the two incidents, even though they happened eight months apart. Randolph eyes him suspiciously and asks, "Donavan had shoes on when you chased him?"

"Ummm," Sean pauses and looks at his Gatorade. He appears trapped by Randolph's simple question. There's an awkward silence as Sean searches for an answer. He then begins a meandering explanation. "Yeah. If those are even his shoes, because I don't know."

What the fuck? He already told us he found Donavan's shoes, and he's proudly flashing pictures of the mud-soaked shoes around my RV. But when Randolph asks if Donavan wore shoes when Sean and the gang chased him into the forest, Sean suddenly doesn't know who owns the shoes that are pictured on his phone? A second ago he was proud of the pictures of Donavan's shoes, and now he doesn't know who they belong to?

I'm no detective, but when Sean speaks, I believe we're hearing a murder confession in jig-saw form. His speech pattern is unnaturally slow. He's putting way too much thought into simple answers, and his emotions are all over the place.

"What was the weather like when you chased him?" I ask.

Sean starts to describe the fog and snow from October 9, but suddenly bursts into tears. That makes no sense. The weather is not an emotional question. The concerned look on Renae's face tells me I'm not the only one noticing Sean's emotional outbursts to simple questions.

I'm trying not to put too much pressure on Sean. Whatever he did to Donavan, I'm just glad he's here helping Donavan's family. This is what I wanted. It's what we prayed for last night.

"I don't like it down there," Sean confides to me. "It gives me the creeps. I call it the Valley of the Shadow of Death..."

"What?" I can't believe Sean chose those words.

"Yeah. Some of my guys are real spooked," he continues. "One of them threw up."

I've never heard hotel grounds described in such a Biblical way. And why would Sean call it "The Valley of the Shadow of Death" unless he knows people died there?

One thing I know is I wouldn't want Sean chasing me into the woods with his machine gun hat and serrated knife. There's an army of guys like Sean at the school bus stop. They're the reason I carry a gun.

Word on the street is that several Storybook workers quit this morning because of what we found behind their building. I wouldn't quit my job because of a mistake or accident, but I'd quit if my company got caught murdering customers and hiding them in the woods.

No paycheck is worth this soul-sucking nightmare. My neighbors are right to be afraid of that God-forsaken "hotel." In 2013, I reviewed Storybook Inn and Above it All Treatment on www.Yelp.com. You can read my reviews online, and you can count this chapter as my two-year update.

Yelp says the hotel is "open now."

If you look at the Yelp pictures, you'll see the balcony and the stairs where Donavan fled for his life. And according to Janice Rutherford, you can book an overnight stay. But I gave it one star, because I suspect some people never checked out.

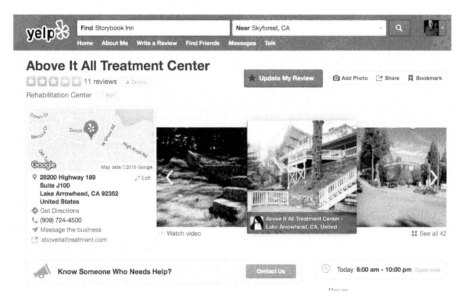

IF THE SHOE FITS...

Donavan's mom and her friend Renae listen suspiciously to the man who hunted Donavan the day he disappeared. After eight torturous months of no information except, "Donavan left. We don't know where he went..." Sean now magically knows where Donavan's shoes are.

I understand the suspicious look on Renae's face. How does Sean know those are Donavan's shoes and sweatshirt unless he was there when Donavan lost them?

I hate to bother Cyndi with morbid questions, but someone will have to go behind the hotel and identify Donavan's shoes. "What size are Donavan's shoes?" I ask.

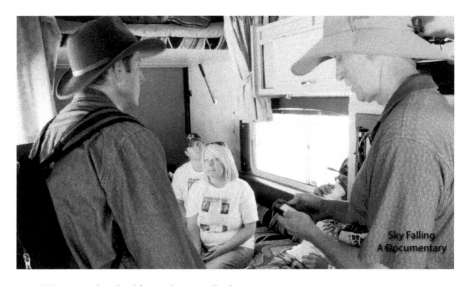

"Ten and a half to eleven," she says.

God bless this poor family. The sheriffs aren't enforcing the law here, so I have to do something I'm not comfortable with.

"OK." I nod. "Shoes, size ten and a half to eleven."

I'm already at my wit's end. I turn to the only professional among us and ask Randolph, "Do I need to write that down?"

"No. I got it," Randolph says, as he taps into his iPhone. Thank God Randolph is here.

Cyndi and Renae describe Donavan's sweatshirt to Randolph, who continues taking notes. "It's an extra-large sweatshirt with a Nike Swish." I'm saddened because they're describing the sweatshirt in the picture on Sean's phone.

The situation is getting worse. My crazy amateur search is now a murder investigation. Have you seen how long it takes me to build a barbecue grill? I shouldn't be in charge of anything like this. The hotel, with hundreds of culprits and potentially thousands of victims, needs a federal law enforcement task force. Not me with my quadcopter sticking out of a half-zipped backpack.

I wonder if the reason the sheriffs aren't here today has anything to do with Janice Rutherford's re-election. At first, I was shocked when the sheriffs didn't show up this morning. But on the other hand, Janice has allowed Storybook Inn to run illegally at a school bus stop for more than a year. Maybe crimes at Storybook Inn will remain unsolved until after the election.

ANGEL FLIGHT

CRAIG PULLS ON HIS CYCLIC stick and the helicopter rolls into a left-hand turn around Storybook Inn so Sam can take pictures of the hotel.

I wonder what goes through Sam's mind as he photographs a creepy van pulling into Storybook Inn with a fresh load of addicts just like Donavan.

You might think a hundred thousand dollars would get you a ride in a nice new van, but some of the vans in Sam's viewfinder are missing hubcaps. The beater vans have curious dents and scratches. As Sam clicks the shutter, the van doors pop open, and sick customers step into a medical facility that authorities say doesn't exist. Meanwhile there's a dead body in the woods behind the building.

Sky Falling - A Documentary

Six months before Donavan set foot in Storybook Inn, we asked Janice Rutherford to let law enforcement shut down Storybook Inn. Janice refused, insisting that it's just a "hotel." But when Suzee and I tried to check into the hotel, World War III erupted. Curious. Even more curious is the hundreds of pages Suzee received from California state regulators who declare that Storybook Inn is not a hotel, but an illegal rehab.

Re: Complaint Number 12-251U

The Department of Health Care Services (DHCS), Substance Use Disorder Compliance Division (SUDCD) performed an on-site investigation into complaint number 12-251U concerning an alleged unlicensed facility operated by Above It All Treatment Centers (AIA) located at 28717 California 18, Skyforest, California 92382. Results of the investigation determined AIA was operating an alcoholism or drug abuse recovery or treatment facility without a license in violation of Section 11834.30 of the California Health and Safety Code and Section 10505 of Chapter 5, Title 9 of the California Code of Regulations (CCR).

Health and Safety Code Section 11834.30 and Section 10505, CCR, prohibit operating, establishing, managing, conducting or maintaining an alcoholism or drug abuse recovery or treatment facility in California without first obtaining a current, valid license.

On January 14, 2014, DHCS, SUDCD issued AIA a Notice of Operation in Violation of Law, ordering AIA to cease providing all alcoholism or drug abuse recovery or treatment services at the address noted above within 15 days of receipt of the notice. AIA provided DHCS, SUDCD with written verification that all alcoholism or drug abuse recovery or treatment services had ceased as of January 21, 2014. Please inform DHCS, SUDCD if your office is made aware of services being provided at the address noted above that require DHCS licensure.

State records indicate that the Storybook Inn is owned by a construction company that the sheriff captain said is run entirely by criminals.

"How do you go from construction to drug rehab?"

— Shannon Doyle, Donavan's dad

The slick ads are impressively polished but contain pictures of the same "hotel" that our helicopter is hovering over.

If you think it sounds like a horror movie, you're right. But this is real. And I think it's what Randolph Beasley was referring to when he called Storybook Inn a "Recipe for Disaster."

Last year I warned County Supervisor Rutherford. I said, "Mark my words. This will get worse." And now here we are, searching for a body behind the building.

I hope you're reading this, Janice. I recorded the meeting where you refused to stop these criminals. Copies of the video were sent to a documentary producer in New York. Not even my death will prevent this from going public.

The bad guys repeat Janice's talking points in person and in mysterious phone calls to my home. They say I should keep quiet, and they know where I live. I've been told to stop talking at San Bernardino County public meetings, and two people have told me to take Janice Rutherford out of the documentary. I suppose "don't fuck with us" goes all the way to the Supervisor's office.

I know it's dangerous to be seen fucking with the "don't fuck with us" gang, but they're endangering children at our school bus stop. So I had to make a choice, and I chose not to keep quiet about Storybook Inn. I wish Donavan's family had heard my warning before they sent their son in there.

October 9, 2013 was a school day. Children were there the day Donavan fled for his life from the school bus stop.

As Craig hovers, the chopping sound of his blades thunder across the mountain. None can escape it. That sound sends a message. "The victims of Storybook Inn are not forgotten. We will honor them."

I press the key on my Icom VHF radio and ask Craig if he wants to land in San Bernardino and wait for us to climb into the Valley of the Shadow of Death. Craig says he'll land and turn the engine off to save fuel. He'll wait for to us to call him in to search for Donavan's body.

But Craig has bad news. We're already out of flight time. The credit card I gave to the helicopter company is maxed out.

I was hoping a sheriff's helicopter would come and fly with Craig today, but so far the sheriffs haven't even sent a foot patrol. Craig and Sam are the only air support we have.

We're out of flight time, and we still have more searching to do for Donavan's family. Oh well. I'll have to come up with the money later. So I text Craig to keep flying until we finish the search. Without being asked, Craig texts back that he'll pay for the fuel we'll need to complete the mission. He didn't have to do that. God bless Craig and Sam Dyer.

NO VACANCY

As I get near the Storybook Inn, gangsters step out. Their wicked tattoos and death-metal T-shirts don't match professional uniforms I've seen at hotels or medical facilities. Gangsters shoot murderous looks at me. The sign says "No Vacancy." I'm not surprised they're full to capacity. Every day they check in more patients from all over the country who probably don't know much more than the beautiful pictures of Sky Forest they saw on the website. It looks like Betty Ford on the website—a first class rehab. But these guys look more like a street gang than recovery counselors.

I estimate 1,000 drug addicts enter Storybook Inn each year, where a full treatment costs over $100,000. That's a lot of money. So it's a shame when I have to borrow money to hire a helicopter to search for their patients.

The helicopter bill is piling up on my credit card, but the cost would be nothing to a business that's billing millions of dollars each year. They don't even have the usual costs of running a business. According to eyewitnesses and county officials, the Storybook owners didn't have to pay for county licenses and fees like Santa's Village and other businesses in Sky Forest. One county worker told me the illegal rehab doesn't even have an occupancy permit.

Sources say that gang members earn $8 to $15 per hour. I believe that's less than what licensed caregivers are paid at a licensed medical facility. I'm no genius, but it looks to me like management knows what they're doing. I wonder if they've known all along what happened to Donavan Doyle.

Search volunteer Gene Barnharst stares in wide-eyed

amazement at Sean's pictures, but Hugh Campbell eyes them with disgust. Hugh is not surprised that a missing man's clothes and shoes are hidden in the woods behind Storybook Inn. As President of the San Bernardino Anti-Corruption Team, he knows how far criminals and corrupt politicians will go to keep a secret.

Hugh didn't buy the stories that came out of Storybook Inn when Donavan disappeared.

"How can you have so many conflicting stories unless everybody's lying?" Hugh said. And as usual, Hugh was right. For eight months the hotel staff pointed in every direction except behind their building.

As the hulking hotel worker thumbs through pictures of Donavan's things, Hugh watches him like a hawk. God bless Hugh Campbell. He knows how corrupt San Bernardino County is, but he's willing to pitch in to help our town and give something back to Donavan's family.

When Donavan's dad sees pictures of his son's shoes and clothing, he bursts into tears. This is hard to watch. God bless Shannon Doyle. If Donavan were still alive, he'd be wearing the items in Sean's pictures. Any glimmer of hope is wiped away as he swipes through Sean's camera roll.

Workers at the school bus stop stare at us as an awful scene plays out in front of their illegal business. The parking lot is full of the dreaded vans that are used to chase down "escaping addicts" like Donavan.

Wendy McEntyre puts her arms around Donavan's dad and comforts him. Wendy also lost her son at a drug rehab. She knows exactly how it feels and can help Donavan's parents in a way no one else can. God bless Wendy for coming out to help a stranger who's suffering just like she is.

HELP FROM ABOVE

H<small>UGH PREPARES OUR VOLUNTEER SEARCH</small> team for the push to Storybook Inn.
   Craig is our eye in the sky.

My eye in the sky is Nathan's plastic toy, but as usual, Nathan was right and the toy is working perfectly.

Randolph is handling crime scene investigation for us. Whatever we find today, it will be handled by someone who cares enough to do the right thing.

Wendy, the Pasquales, and my neighbors are counseling victims. Some of my timid neighbors who live in fear of the "don't fuck with us" gang are coming out of their homes to offer support for Donavan's family. They're bringing food and hugs to our search effort.

I'm not sure how to describe Robert Larivee. I've never seen so much good humor and knowledge in the same shoeless person.

And Robert is also a minister, who brought church to the scene of the crime.

Meanwhile, Gianny and I are the only media. The "real media" knows about Janice Rutherford's school bus stop scam, but they buried the story. They ignored all my phone calls and emails. That's how I learned that big media companies will hide the truth to protect powerful politicians, even at a school bus stop.

So Gianny and I must document everything ourselves. Sweat pours off Gianny's brow as he rapidly switches between his lightweight DSLR and his Red Cinema camera.

Our volunteers brought a sense of order and purpose to a place where the government isn't governing, counselors aren't counseling, and reporters aren't reporting.

Our volunteers are taking care of everything that needs to be done for Donavan's family. And that's a good thing. I wish I could take credit, but I couldn't even get the search invitations printed correctly. My cockamamie shoot-from-the-hip planning is not the reason that Donavan's family has everything they need right now. Someone I can't see is helping us.

Someone I can't see is calling the shots and talking me though each step. It's the voice I heard when Donavan's mom and dad came to my house. It reminds me of the voice I heard at the Calgary Airport when Gibbs sent information about my child custody case a week after his death. I started the search this morning with the voice of someone I can't see. But now we're finding things. I can tell by the look on our crime scene investigator's face that we're finding things with unnatural speed. I don't know what we're going to find next, but I know we're being helped and pointed in the right direction.

ROBERT'S PLAN

I NORMALLY CAN'T TELL WHEN PEOPLE are lying to me, but with Sean, I think I can tell. I don't buy Sean's story. It doesn't add up. His timeline and emotions are contrary to common sense.

Robert Larivee pulls me aside and whispers, "I think he's hiding something." Sean scratches his head and thinks carefully before answering Randolph's question about Donavan's disappearance. Robert whispers, "I asked Sean the same question and got a different answer."

Robert points to his map of Sky Forest and says, "I think we should send a team with Sean behind the hotel, and another team should go back to Sky Forest Springs. One of the van drivers said they hunted Donavan below the springs. We can search that area and circle up to Storybook Inn." He draws a circle with his finger over the area on the map that Sean referred to as the "Valley of the Shadow of Death." Uh oh. Robert circled the place where Craig said we would find Donavan.

I'm worried about what we're going to find in that circle, but I like Robert's plan. We'll pull our search team from Switzer Park and send them with Sean to Donavan's shoes. Randolph will investigate the crime scene, and Gianny will document with his camera. Unfortunately, Donavan's dad will be with them. He'll have to enter the "Valley of the Shadow of Death" where his son's body has likely been hidden for the past eight months. That makes me nervous. But Donavan's dad is the only member of our team who can confirm the identity of Donavan's shoes and clothing. So he'll have to be there when Randolph examines them.

A father should never be allowed near a crime scene where

his son's decomposing body might be hidden. This is a job for the sheriffs. But the sheriffs aren't here. So we have to do this ourselves. God be with Shannon Doyle. The poor man is already grieving.

I take a nervous breath. I count the heads of our volunteer team and say, "OK. Let's go."

SNAKES

It's strange that the sheriffs say they've never been to Storybook Inn. And yet the Storybook staff calls their backyard the "Valley of the Shadow of Death." Now I understand why Donavan's family came to me for help in the first place. Who will help the victims if the sheriffs are covering up the crimes?

And now I'm afraid we're crossing dangerous lines. The sheriffs are on the criminal side. That's why some of my neighbors stopped talking to me. Any one of us could end up murdered and dragged into the woods behind the hotel.

We approach the cliff. My stomach rises uncomfortably into my chest as I peer over the vertical drop at Sky Forest Springs. This should be done by professionals, but the professionals aren't here. Instead of professionals, I have what Randolph calls "a Motley Crew" of amateurs.

As I slide my feet over the precipice, they dangle in mid-air. I'm trying to get a foothold, but there's no place to put my feet. And guess where I am? The EXACT place from my Tom Cruise falling nightmare.

In my dream, Tom Cruise tells me and the neighbors how to do a base jump from this same spot. He points to the abyss where my feet are now dangling and says, "You guys jump in here!" But in my dream, Tom Cruise has a parachute and I do not.

"Snake!" Back in the waking world, my neighbor points to the biggest rattlesnake I've ever seen. The viper coils to strike. I'm afraid to breathe as the venomous reptile gives me the evil eye. I'm clinging to roots, vines, anything I can, but the root breaks and dirt flies as I make circles in mid-air with my right hand. I glance down at the snake. My frantic movements will no doubt cause the snake to strike. But he doesn't bite. He just points his head at me with his body spring-loaded like a venomous cannon. There's no easy way down this cliff, and there's no safe way around the snake. I slide awkwardly as the rattlesnake tracks me with unblinking eyes.

I pause for a moment and take in the sight of my neighbors and friends. We're amateurs entering an unforgiving realm of life and death. Each stumbling step pulls us away from the safety of everyday life. Now gravity and our own choice to help a lost family pulls us toward the "Valley of the Shadow of Death."

INTO THE VALLEY OF THE SHADOW OF DEATH

MEANWHILE, AT THE "HOTEL..."
Gianny walks through the flowers where he filmed Wyatt back in February. Gianny and I had to run and hide the documentary camera from gangsters. Now Gianny is back in the same flower bed with his camera searching for a missing "hotel guest."

Donavan's dad points to the spot where he found his son's belongings. After Donavan disappeared from the hotel, he found Donavan's things tossed out on the ground. Donavan's clothes were soaked in urine, a tell-tale sign of foul play.

Donavan's Aunt Cheri spoke at the Candle Light Prayer Vigil, and she complained bitterly, "Why didn't they pick up Donavan's things and give them to his family? He is a human being!"

Janice Rutherford said that Storybook Inn is just a hotel. But what kind of hotel workers fight with a guest and then throw his belongings out the door and leave them soaked with urine?

Aunt Cheri pleads, "I know they see his things there. Why don't they pick them up and give them to his family?" Aunt Cheri doesn't know it, but she's asking why a street gang won't lift a finger to help us. Aunt Cheri doesn't know them like I do. She didn't have to hide from the gang as I did when I searched for Donavan on Mother's Day. I had to be very careful not to let the gang see me searching alone in the forest where they may have hidden Donavan's body.

Donavan couldn't have known when he checked into the Storybook Inn that he was crossing a point of no return. Donavan and his family didn't know the Storybook Inn like I do. Maybe it's because county spokesman David Wert said the hotel is "totally in compliance with all state and county regulations."

In a bizarre twist, Gianny now follows Sean through the flowers to the dreaded "hotel." If you put a gun to my neighbors' heads, I don't think they would follow Sean there. My neighbors aren't fooled by the happy facade and nursery rhyme paint scheme.

Creepy vans line the front entrance of the hotel. The rehab is bringing in millions of dollars, but for some reason, the vans are missing parts. I hope the dents are from hitting a tree or another van, but with this gang, I don't know. There may be blood on them.

The bedraggled vans have become an ominous presence in Sky Forest.

The vans are everywhere, with thugs on board watching us. They hunt anyone who escapes the hotel. One hotel guest hid under my neighbor's stairs. When questioned, the frightened man said he was hiding from guys in vans who wouldn't let him leave the hotel. According to Sean's statements, he's one of the guys.

When Gavin Brennan and David Mickler tried to film the school bus, hotel vans encircled them, with gangsters shouting threats. "We're watching you!" one thug yelled.

It happened to me, too. Sometimes they yell threats as they roar past with open windows. Sometimes they pull over and yell. In these moments, I have to consider whether to walk back to my house or go someplace else for a while so the thugs can't see which house I live in—because Wyatt lives there too. And I have to try to protect him from the violence that causes people like Donavan to disappear.

SKY FALLING
A Documentary

The vans menaced Donavan's family when they came to search for their missing son. As Donavan's dad noted, "There are big guys in vans, watching us..."

Donavan's dad confided to me, "Now I hate vans. I get angry any time I see a van anywhere..." That sounds weird, to hate vans. But I know how he feels. I've lived with these vans for over a year. A chill goes up my spine whenever I see a cargo van with rows of seats.

I don't know how this will end. But when Donavan's mom and dad leave town at the end of our search, Wyatt and I will still be surrounded by vans and gangsters...and so will the school children...

Volunteers hurry to keep up as Sean leads them past the vans into the secretive realm of the "don't fuck with us" gang. Gianny glances at the vacancy sign hanging over the front door of the strangest hotel in California. Like Hansel and Gretel, Gianny and Randolph find a cheerfully painted facade. But the air at Storybook Inn is thick with tension. Donavan's Aunt Cheri says walking into this hotel is like a scene from a horror movie.

Gianny knows they hate cameras at the illegal business. Imagine carrying a camera into a mafia stronghold. Gianny grips his camera tightly as Sean guides them into the hotel property.

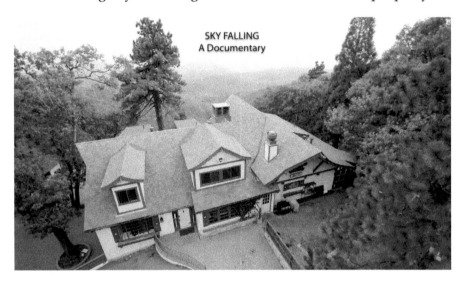

SKY FALLING
A Documentary

Hugh Campbell is the founder of the San Bernardino Anti-Corruption Team and the leader of our ground search expedition. His eyes grow wide as Sean leads them around the hotel toward Donavan's shoes. Gang members shoot murderous looks as volunteers follow Sean to the balcony that overlooks the backyard.

Hugh pauses on the balcony and takes in the view that you can see if you go to www.aboveitalltreatment.com.

The pictures look great on the web site. The cheerful slogans sound nice. But if the web picture was taken between December 2012 and December 2016, there may be multiple dead bodies in the picture. I know it's hard to believe, but our knife-wielding hotel tour guide calls the view from their website the "Valley of the Shadow of Death." And I believe him.

Sean leads our searchers down the stairs that Donavan jumped over in an attempt to escape from the hotel.

Our primary goal at Above It All Treatment is to provide our clients with the tools and structures they need to embrace lifelong recovery, and to maintain sobriety long after they leave our treatment center. We combine proven therapies with cutting-edge techniques, seeking to provide clients with a path toward independence and good health.

Our team members bring a variety of expertise—but more than that, our staff members are passionate about the well-being of every single client, and we let clients know just how much they are cared for and supported from the second they arrive to well after they leave.

CALL NOW: **877-574-0177**

SKY FALLING
A Documentary

Just like the kid we found hiding under my neighbor's stairs, Donavan wanted to call home and have his family rescue him. My neighbor gave the kid she found a phone, and he was able to call his family to come to Sky Forest and rescue him. But no one helped Donavan.

Another addict from New York told me, "They wouldn't let me leave, so I ran for it between cars on the highway." He evaded the "don't fuck with us" gang by walking 19 miles down a busy road. His frantic dash in traffic lanes with cars weaving around him was like a scene from a Quentin Tarantino action movie. But at least he got away. He made it out alive.

Escaping hotel guests have to dodge high-speed cars while trying to run away. I've seen drug addicts running this way dragging heavy suitcases. They're in mortal danger. But every step in high-speed traffic gets them further away from the Valley of the Shadow of Death.

Gianny carefully hides his camera from the Storybook thugs.

As they exit the hotel, he readies his camera and focuses his DSLR lens on volunteers who slide down the slippery slope behind Storybook Inn. The knife-wielding rehab worker leads them straight down behind the hotel to the dead trees where I crashed my drone. Larry Pasquale loses his footing and careens head over heels. He jams his fingers into the mountainside and claws to a halt in front of a massive fallen storybook pine. Here he must crawl like an army recruit to get under the fallen tree and enter the "Valley of the Shadow of Death."

One at a time our volunteers crawl through the barrier of dead trees, with rotting wood and dead branches reaching toward them like wooden skeletons. Randolph grabs a dead branch to stop himself, but it breaks and sends him into a downward tumble. Dead trees offer no comfort here. Randolph carefully jams his boots into the rocky terrain to stop his slide into the abyss of my nightmare.

Meanwhile, Sean climbs into the valley with ease, as though he's been here many times. Where volunteers stumble and slide, Sean moves quickly, with a familiar knowledge of where to place his hands and feet.

Gianny takes a deep breath of hot mountain air and thinks about the suspicious coincidence at Storybook Inn. "What's the chances of them suddenly finding Donavan's shoes when they've said they know nothing…for eight months?"

Gianny is right. After eight months of claiming not to know where Donavan went, Sean suddenly leads our volunteers into the woods behind his building.

For nineteen years Randolph investigated crime scenes, but he's never seen one like this. Normally Randolph is surrounded by deputies and detectives trying to solve the crime. But there are no deputies or detectives in the Valley of the Shadow of Death, just me and my friends.

Sean moves so quickly down the Storybook slope that only the fastest volunteers can keep up with him. Hugh is beside him when he lands at the bottom of the cliff.

Hugh looks around. Something's not right. Sean points to Donavan's shoes and sweatshirt. But how did Sean find these small items in a vast forest so quickly? Hugh eyes the knife-wielding hotel worker suspiciously.

High above them Donavan's dad slides down the cliff, loosing rocks along the way. He grabs a tree root to stop his slide as an arsenal of falling rocks rain down on Hugh and Sean.

"Are you OK?" asks Larry Pasquale.

"Yeah, I'm OK," says Shannon Doyle, gasping for air.

"Where'd they go?" Larry asks.

Shannon looks down the cliff below their feet as the rocks fall toward Sean and Hugh.

"Straight down."

Gianny and Larry keep a worried eye on Shannon. Everyone is worried about what we'll find in the valley below his feet.

The rocks under Larry's feet give way, sending him like a human torpedo down the cliffside.

"Look out down there!" Gianny yells to Hugh and Sean.

Sean and Hugh look up to find a fusillade of falling rocks, with Larry coming after. Hugh and Sean dive for cover from the falling rocks as Larry digs in with hands and feet. He grabs a hold and stops his fall. He looks to his left and sees something that makes him yell, "Wallet!"

Shannon Doyle climbs down beside him and looks at the wallet. With tears in his eyes, he nods. It's Donavan's wallet.

We really should have a homicide detective, but we have a cameraman. Gianny focuses on the wallet as Randolph snaps on a pair of latex gloves.

Gianny zooms his Canon lens as Janice Rutherford's opponent in the coming election looks inside the wallet of a missing Storybook Inn guest. There's no money in the wallet, just a piece of paper with a handwritten name and phone number. Randolph reads the name and number out-loud as Gianny records.

SKY FALLING
A Documentary

Shannon Doyle watches as our volunteer CSI investigator examines his son's wallet.

One by one, the volunteer searchers climb down to Donavan's wallet. Gene Barnharst wipes sand and dust off his forehead after slip-sliding down from Storybook Inn. He pauses to catch his breath and glances at Wemyss, who has barely broken a sweat. Wemyss quickly surveys the crime scene as the hotel worker points further down the valley. Sean says they should continue on, suggesting that there's more to see. Wemyss surveys Donavan's shoes, sweatshirt, and wallet. Hugh looks carefully at the knife-wielding rehab worker and whispers to Gianny, "Sean knows more than he's saying." Sean starts marching and leads deeper into the Valley of the Shadow of Death. Wemyss volunteers to follow him.

But, Randolph isn't done with the shoes and sweatshirt. He extends a latex-gloved hand toward a muddy shoe and looks inside for the size tag.

The scene looks like an episode of CSI, except that there are no law enforcement officers to help Shannon. There's no crime scene tape. There are no boundaries between a homicide and the victim's father. Larry takes a long, worried look at Shannon.

Larry supports his friend, but he doesn't want Shannon to take another step into the "Valley of the Shadow of Death." He grabs Shannon's shoulder and says, "Hey. I say we look at (the clothes and wallet) and then let's go. I don't want you to be here if they find anything else…"

God bless Larry Pasquale. He's the best kind of friend. And he seems to know where Sean is leading the searchers next. Larry doesn't want Shannon to stumble across the decomposing body of his son.

THE ROAD TO NOWHERE

Hugh warned me about stinging nettles on our pre-search scout mission. Now I'm getting a good look at them as Robert and I take turns stepping over a rattlesnake. "Please Lord, let me get out of here without a snake bite."

They say rattlesnakes can strike more than the length of their body. I hope that's not true. I'm much closer than the length of the snake at my feet.

My snake dance is interrupted by Randolph calling on the walkie, "CSI to Robert, we found a wallet that Shannon has identified as the wallet he bought for Donavan." I guess I can't do two things at once, because I fumble the walkie and lose my footing. The good news is, I'm not falling onto the snake. The bad news is, I'm falling face-first into the stinging nettles. Ouch! My chin pulsates as if a swarm of bees is poking me with stingers.

Embarrassed, I climb out of the stinging nettles, rubbing my chin, which feels like it's on fire. Everyone stops in their tracks to hear walkies as Randolph describes the shoes and sweatshirt to Robert. Robert has a look that says something isn't right. He pulls out a terrain map and looks at where we are. We're in the canyon where a Storybook van driver told us to look. But there is nothing here except stinging nettles and rattlesnakes. Meanwhile, Sean walked through the hotel to Donavan's wallet, shoes, and sweatshirt.

So why are we on this wild goose chase? If Donavan's things are behind the hotel, why aren't we over there? If Storybook people know where Donavan's things are, why didn't they tell us for eight months? And why did they tell us to look where the snakes and stinging weeds are?

FOUL PLAY

Behind Storybook Inn, Wemyss notices the trail ahead has been trampled down. Sean says they're searching new ground, but it looks like someone has already walked here. Wemyss looks closely at the tracks in front of him. They start at the Storybook Inn and lead to Sean's feet.

The sneaky footprints give the scenic canyon a sinister vibe. Sean follows the footprints but suddenly stops. Water from a small stream flows through the pile of wood. Wemyss' eye locks onto something Randolph taught us about in his CSI class. He turns to Sean, but Sean's not there. A second ago Sean was with Wemyss, but as he arrives at the pile of sticks, Sean has suddenly disappeared. Wemyss stands alone in the canyon. His sixteen-year-old eyes glimpse a gruesome but unmistakable human form. He takes a deep breath and presses the button on his Motorola walkie.

Across the mountain walkies crackle with the sound of Wemyss' voice, "We uh…we found the body…"

Randolph freezes. He glances nervously to Donavan's dad and family friend Larry Pasquale, who fill the air with the sound of heartbreak. Larry speaks between muffled tears, "Oh no! Oh no! Goddamn it…"

Randolph seems not to know what to do. He's investigated hundreds of murders, but he's never had the victim's father standing next to him. Randolph nervously keys his walkie, "Copy that."

"FUCK!" the canyon echoes with sound of furious anger and grief. Shannon puts his head in his hands and cries the most painful tears a parent can shed. It's a heartbreaking, soul-sucking sadness that's hard to watch.

For a moment, Gianny films our volunteer CSI investigator with the father of the victim, who wipes tears from his eyes. But Shannon's tears won't stop coming. Gianny is an expert filmmaker, but he's also a decent human being. He flicks a switch with his thumb and turns his camera off. We filmed Shannon Doyle enough today. Gianny lowers his camera and allows Shannon to grieve for his son in peace.

Randolph gathers himself and keys his walkie again. "What is your location?"

Wemyss answers back, "about four hundred yards from the shoes and sweatshirt." Randolph lowers his walkie and gazes into the Valley of the Shadow of Death. Donavan's body is around the next bend.

In some ways, it's a miracle. The Doyle family would have lived their whole lives without knowing what happened to their son. At least now they know the truth. And instead of searching and praying and worrying for the rest of their lives, they can finally lay their son to rest.

But knowing that Donavan was murdered and dumped behind the hotel doesn't give me peace. Donavan's body, broken and discarded from Storybook Inn, is what I've been warning people about since Storybook Inn opened. And I'm not the only one. When Randolph Beasley learned what they were doing at our school bus stop, he called it "a recipe for disaster." And now Randolph will have to snap on his latex gloves one more time in the Valley of the Shadow of Death to examine the body of Donavan Doyle.

Back in the valley of rattlesnakes and stinging nettles, Robert looks as sad as I feel. We're both fathers. The news of Donavan's body is not what we want to hear.

You might think that the discovery of a dead body would be the end of Storybook Inn. But in a county ruled by crime and corruption, this story is just beginning. Storybook Inn will remain open.

The sheriffs won't investigate Donavan Doyle's murder. They're not asking questions in the Valley of the Shadow of Death. In fact, they're going to leave parts of Donavan's body behind the hotel for another three months. If you think that sounds impossible, then fasten your seatbelt. My next book will take you on a wild ride.

As Gianny focuses his world-class looking glass on crime in the woods behind the hotel, some of the crimes are committed by sheriffs. Charles is tracking the story from New York. He's saddened by what we found today, but eager to tell the story to his documentary audience. We're being threatened with lawsuits by the bad guys, but Charles won't quit until the general public sees the footage from Gianny's camera.

I'll continue writing about my life in the sky and my fight to free my town from organized crime. In my next book, we'll take a look at Donavan's body, left lying in the woods behind Storybook Inn for eleven months. Sheriffs are required by law to investigate, but they're not following the law here. Prepare to be shocked.

You won't find the story in San Bernardino County archives. According to San Bernardino County, the business Donavan's family paid into doesn't exist. After four years, I can only find a few county documents that even mention the names Donavan Doyle or Storybook Inn. Other victims aren't named at all. They aren't known. Those people simply disappeared from a hotel for which there is no license, no permit, and no records for a single hotel guest.

RECORD BREAKING

THE PEOPLE INVOLVED WITH STORYBOOK Inn have been working hard to keep this story a secret, and in my opinion, they're not getting the recognition they deserve. Since there are no records at the hotel, I'd like to close this book by recognizing their achievements:

Although we informed County Supervisor Janice Rutherford, the Storybook Inn got a record-breaking fast approval granting a drug rehab permission to operate at a school bus stop without hearings, signs, licenses, or permits. You might think we just missed the signs. But with school children gathering six times a day, we'd have noticed signs announcing a business that has already been thrown out of other towns for criminal activity.

Rutherford's county government also set a record for the slowest county approval right next door! The owners of Santa's Village have been waiting three years, even though they posted signs, held the required public hearings, and got licenses and permits. Santa's Village got health inspections, plumbing inspections, environmental impact studies, traffic and parking studies, but still can't open. Meanwhile, Storybook Inn welcomes thousands of guests without any of those things.

Home ▸ News ▸ Subscription News

Santa's Village Delayed

Story   Comments   Share   Print   Font Size:

Tweet

Bill Johnson, project manager of SkyPark at Santa's Village, updated mountain contractors on the park's progress.

Posted: Thursday, March 3, 2016 6:00 am

By Mary-Justine Lanyon, Editor | 0 comments

Bill Johnson's association with Santa's Village goes back to when he was 13; he worked at the park for three years.

"Now I'm the head elf," he jokingly told the members of the Association of Building Contractors of the San Bernardino Mountains at their February meeting.

As the project manager for SkyPark at Santa's Village, Johnson said he started the process "thinking we could move in like a remodel because it already existed. We thought the process would be a lot simpler."

"We immediately attacked the project with everything we had," he said. "We cleaned out the old. But then the county came in and said they didn't want us to clean up anymore."

Sky Falling
A Documentary

Owners of a restaurant in Sky Forest spent years but still couldn't get permission to sell a sandwich. County enforcers pushed the restaurant to bankruptcy. Thank God our county government was able to protect us from those sandwiches.

You might wonder, "How can a hotel set so many records, when Santa's Village and the sandwich shop can't open?" I don't know how it works, but regulators, business owners, and residents all say they witnessed Janice Rutherford's personal intervention. It's one of the reasons none of the business owners I spoke to would allow us to hold a candlelight vigil for Donavan. Not even the church would let us pray for a victim of Storybook Inn. People are afraid, so they're careful to not be seen asking questions or helping victims of the scam at Storybook Inn.

In all my years, I've never seen anything like this in the presence of school children. That record takes my breath away, and it's the reason I got involved. The school bus stop is the reason I never quit or backed down when confronted by criminals, killers, crooked officials, or dirty cops.

The hotel's water usage is another incredible record. There are laws that keep a business from using a town's water supply. But when our water company told Janice Rutherford, "We don't have enough water for this," she refused to take action. Then the Storybook gang completely drained our mountain spring.

Since we ran out of water, Sky Forest residents have been paying for water to be delivered to our town. It's a shame for our town to go thirsty when we have our own mountain spring. But Janice knew this would happen. We talked about it.

Todd Pahl of Sky Forest Mutual Water Co. explained to Janice Rutherford how Above it All Treatment empties our water tanks every two days.

Todd also says that the sewage overflow from Storybook Inn is contaminating the town's well at Sky Forest Springs. I'm not surprised the sewage overflowed. The hotel has more residents than the facility can accommodate. Neighbors say they endured nauseating smells from the Storybook Inn for years because the overflowing sewage from the hotel runs down the mountain into the Valley of the Shadow of Death. Meanwhile, more people are crowding into a facility that was made to accommodate a fraction of the gang that lives there.

I'd love to take my son to Santa's Village. But so far the county has it padlocked.

When we asked Janice how she can keep an illegal drug rehab open at our school bus stop, with no zoning approvals, public hearings, or impact studies like the ones at Santa's Village, she told me Above it All Treatment is not a rehab. I don't know how she can say that without changing the company name, business cards, website, nationwide advertising, or medical billing records. Now that people are learning about Janice's record, maybe she can be asked to explain it by someone other than me.

ABOVE IT ALL
TREATMENT CENTER

Call To Speak To A Counselor
888-997-3006

- Home
- > Recovery Programs
- > What We Treat
- > How We Treat
- > Our Rehab Facility
- > Admissions
- > About Above It All
- > Addiction Information
- Blog
- Contact

Need help to free your family from addiction?
We've helped hundreds of families just like yours recover from the pain of addiction

A Recovery Specialist Is Waiting For Your Call.

| OUR PROGRAMS | OUR TREAMENTS | MEET THE STAFF | TOUR FACILITIES |

READ OUR TESTIMONIALS

Standards & Best Practices

The Above It All Treatment Centers drug and alcohol addiction rehab programs blend proven therapies with cutting edge techniques to provide you with successful and lasting treatment.

Our beautiful residential treatment facilities are located in the breathtaking Lake Arrowhead Mountains of California about 2 hours from Los Angeles. Our expert clinical staff coupled with our evidence based treatment programs have been proven to help people gain TOTAL recovery from drugs and alcohol and go on to live happy, healthy lives.

The Above It All Inpatient Rehab Facilities

Get an Email Consultation

Name *

E-Mail Address *

ABOVE IT ALL

(877-941-0879
Non-Admissions Call: 909-337-3366

HOME   DRUG REHAB   ALCOHOL REHAB   TREATMENT PROGRAMS   TREATMENT CENTER   ABOUT   BLOG   CONTACT

TREATMENT CENTER
Expert Care in a Beautiful Mountain Setting

Above It All Drug Treatment Center

All of our treatment programs are tailored to the individual needs of each person entering into our care. Although each drug treatment program is customized, the primary treatment process at our drug treatment center will always include the core therapies as outlined below.

Therapists at Above It All incorporate the 12-step recovery program into all our treatment plans. Thousands of people have found long-term recovery from alcoholism and addiction through this program. The 12 steps serve as a guide to help individuals connect with a higher power of their own understanding, with respect for all beliefs, religious or otherwise. This program also lays the groundwork for participation in a support group, which will help the client long after they have completed rehab.

CONTACT US TODAY
Help Is Available 24/7

Name

Email

Phone

I believe hiring people from "Most Wanted" posters for the school bus stop is another kind of record. With "criminals" threatening to murder us at the school bus stop and law enforcement refusing to help, I'm not surprised that neighbors don't want to go on record and talk about the place. If I didn't write this book, no one outside Sky Forest would ever know.

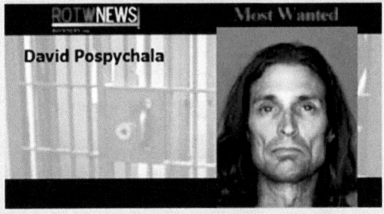

Sources say that Storybook Inn is missing several hundred hotel guests. The hotel quickly out-paced a former record-holder in this category, The Bates Motel. (The Bates Motel is a fictional facade at Universal Studios. According to Janice Rutherford, Storybook Inn is a real hotel, where you can book an overnight stay.) And county spokesman David Wert said the place is totally in compliance with all state and county regulations.

I spoke to people who went to Storybook Inn looking for a hotel room, but instead found what they described as a scene from a horror movie.

Storybook workers tell me they sometimes have twenty tourists a day trying to check in…I hope those tourists are still alive.

Sky Falling
A Documentary

How about the sheriffs' record-breaking murder investigation — completed without questioning witnesses or suspects. In fact, they didn't even send a detective or a coroner to look at Donavan's body. I can't find a previous record where police left civilians in charge of a homicide.

When Wemyss found Donavan's body hidden behind the hotel, his teeth were smashed.

But, in my next book, you'll see how sheriffs removed the body at midnight so the news media couldn't report on it. Fortunately, Gianny set his own world record for sneakiest cameraman by filming the body and sheriffs who snuck into the woods and removed it without investigating the homicide. Does that sound impossible? Wait until you see the documentary.

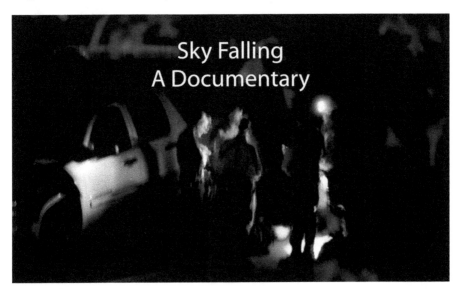

Editors in Hollywood and New York have received twenty-four terabytes of Ultra-Hi-Def images of this record-breaking drama in the dark.

Perhaps it's a world record for law enforcement not to investigate a missing person case for eight months, and then to recover the victim's body in a few minutes at midnight without the use of detectives, forensic investigators, or even lights?

It may surprise the sheriffs to know that I was watching them and recording their achievements with the help of some amazing friends.

Janice Rutherford's reelection was held 14 days after we found the body behind Storybook Inn. Imagine what an interesting record that would be, days before voters pulled the lever for county supervisor? Maybe that's why the sheriffs broke so many records at midnight on Memorial Day, 2014.

But all was recorded by our cameras as it happened. Perhaps our documentary will break some records as well.

The sheriff's department declared that there was no foul play in Donavan's Death without seeing the body, the broken teeth, the crime scene, or the suspects, and before the body was even identified. This must be a new record for law enforcement.

Many people have disappeared from Storybook Inn. But somehow sheriffs knew there was no foul play before we even knew which hotel guest we had found decomposing behind the building. I really believe no ordinary sheriffs could reach that conclusion.

COMING UP

"READY TO MAKE SOME NOISE?" Aaron Fitzgerald asks before pressing the start button on the Airbus AS350. "Ready!" I shout, fastening my seatbelt and plugging in my Gentex helmet.

After three years of corruption, intimidation, burglaries, and murder, it takes my breath away to receive a gift of kindness from Aaron and our friends at Angel City Air. Aaron pushes the button and the battery-powered starter spins the compressor blades to a low whine. He advances the throttle, sending jet fuel into the hot section of the 1D1 turbine. I can hear a ticking sound above my head as lightning from the battery shoots across the combustion chamber.

Tick-Tick-Tick-Boom! Compressed air and gas explode inside the 700-horsepower turbine, which screams so loud Aaron can't hear me until he dons a noise-canceling headset.

As the blades start to spin above our heads, Aaron stares intently at the gauges, making sure a perfect ring of fire grows to seven hundred degrees without melting the engine. He pushes the throttle full forward and the main rotor blades spin so fast they become near supersonic. We are now sitting in one of the most powerful machines you can drive.

Aaron pulls the collective and the power of seven hundred horses push us into the sky.

Aaron usually flies for Red-Bull air stunts and Hollywood movies, but not today. Today, Aaron is flying to Sky Forest to help me save my little town.

I still haven't paid off the helicopter bill from Memorial Day. It's still sitting on a maxed-out credit card. And I can't afford this multi-million-dollar Airbus or GSS camera turret. Thank God Aaron and our friends at Angel City Air are donating the flight and equipment.

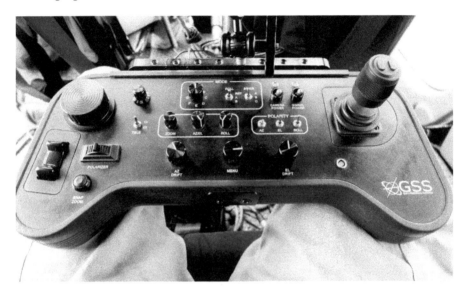

I put the GSS C-520 into stabilized mode, and computers calculate the violent helicopter movement two thousand times a second to stabilize the Red Dragon camera. Rows of crowded LA homes pass beneath the perfectly stabilized Canon lens as Aaron makes the long (and expensive) flight across one of the biggest cities on earth.

Aaron pulls harder on the collective control and the powerful Airbus chops away at six thousand feet of air to push us over the top of the San Bernardino Mountains to Sky Forest.

Machines struggle in the thin air above the clouds. Even the mighty Airbus is gasping for air. The thunderous machine only has two-thirds power here. Aaron carefully harnesses that power with 24 years and 9,000 hours of experience flying for TV and movies to carefully encircle the Storybook Inn.

Normally at a hotel, people smile and wave at a passing helicopter. But the mood here is different than any hotel I've ever seen. One guy has the hood up on a van working on the engine. That's a strange sight for a hotel, and it could explain why people said their cars were sabotaged when they parked nearby.

There's a weird tension at Storybook Inn, as our aircraft thunders above the "don't fuck with us" gang. For three years, the gang has ruled our little town with fear. But not today. Today, the heart-pounding echo of Aaron's helicopter sends a chill down the spine of gangsters who scurry out of sight.

You can run, but you can't hide from GSS, which holds the Ultra-High-Definition camera perfectly still as the Airbus thrashes in the thin mountain air.

My friends at GSS spent ten years and ten thousand hours designing technology to hold the Red Dragon absolutely still so I can capture twenty-four ultra-high-definition pictures per second. Each picture contains nineteen million elements — enough to fill a forty-foot movie screen. 'Ready to make some noise' indeed.

Aaron pushes the cyclic stick, and we move behind the hotel to "The Valley of the Shadow of Death."

"That's where we found Donavan's body," I say as Aaron looks down where the body of a hotel guest was hidden for almost a year.

"Really?" His eyes grow wide.

I doubt Aaron has ever seen a hotel like this. Even the Bates Motel didn't have so many missing guests.

There have been twenty-four deaths that I know of at Storybook Inn, but I can't find county records, so we don't actually know how many people have disappeared from there. We may never know how many were killed and dumped in the forest like Donavan. People generally don't take the time to search for missing drug addicts.

But I took time. I believe that when Donavan was killed, we didn't just lose a drug addict. We lost something precious that cannot be replaced. I hope you enjoy our documentary. It was made with love by me and my friends.

Aaron lowers the collective and the Airbus falls away from the mountain realm of Sky Forest. The wind whistles past my window as we pick up airspeed and start the mile-long descent down to the city.

"Dude! That was insane!" Aaron yells into his headset. "Be careful, bro. Do you have a copy of your footage?"

That's a good question. Everyone asks that. If it weren't for my documentary footage and this book, no one would ever know about the crime and murder at Storybook Inn. And Aaron is right, if I'm the only one with the evidence, that makes me a target for the bad guys.

I'm not out of danger, but I decided not to quit until you know the truth about Storybook Inn. So fasten your seatbelt, because we're about to fly into another chaotic David Alan Arnold adventure. What you saw behind that hotel is just the beginning. In my next book, I'm going to show you what happens when Randolph Beasley snaps on his latex gloves and starts to examine the body of a guest that has been hidden behind the Storybook Inn for eight months. I'll show you what sheriffs are doing in the darkness behind that hotel.

Sheriffs say there are no signs of foul play at Storybook Inn, but I'm going to show you everything. I'll let you get to decide how this story will end.

Imagine what would happen if everyone who reads this book calls their congressmen or the FBI?

I wish this book was better. I wish my grammar were excellent, and the stories flowed like poetry, but I only have a high school education. I prayed for someone smart and rich and powerful to come and save us. But no one came. So I had to get out of my helicopter and write this book myself. It's frustrating and painful, but I have to *become* the man I prayed for. We all will. If the wrongs at Storybook Inn are to be made right, we're going to have to do this ourselves.

I can't promise the bad guys will be stopped, or that justice will be served. But I promise you that I won't quit until the town is safe again.

My name is David Alan Arnold. And this is my continuing story.

If you like the stories in this book, check out the pictures and videos that go with them:

Instagram: @airbornecamera

Twitter: @DavidAlanArnold

Facebook: David Alan Arnold

YouTube: David Alan Arnold
https://www.youtube.com/channel/UChvtDqFS1Uydyl615efR23A?view_as=subscriber

Website: www.DavidAlanArnold.com

COMING UP:

HELP FROM ABOVE

BOOK THREE

Hell to Pay — A True Story

Emmy Award-Winning
Helicopter Cameraman of Deadliest Catch

— **HELP FROM ABOVE BOOK 3** —

HELL TO PAY

A TRUE CRIME STORY

Now A Documentary Series

DAVID ALAN ARNOLD

ABOUT THE AUTHOR

David Alan Arnold is an Emmy-Award-Winning Cinematographer.

He has flown around the world for movies like James Bond and hit TV shows, like Amazing Race, Deadliest Catch, World Series and Super Bowl.

For twenty years, he never talked about his work, until writing this series of books...

www.ingramcontent.com/pod-product-compliance
Lightning Source LLC
Chambersburg PA
CBHW061921030325
22867CB00036B/961